Experimental Methods for Science and Engineering Students

Responding to the developments of the past 20 years, Les Kirkup has thoroughly revised his popular book on experimental methods, while retaining the extensive coverage and practical advice from the first edition. Many topics from that edition remain, including documenting experiments, dealing with measurement uncertainties, understanding the statistical basis of data analysis and reporting the results of experiments. This new edition reflects the burgeoning impact of digital technologies on the way experiments are performed, analysed and presented, and the increased emphasis on the importance of communication skills in reporting the results of experiments. The book is ideally suited to science and engineering students, particularly those new to laboratory or field-based work, who require a coherent and student-friendly introduction to experimental methods. Exercises, worked examples and end-of-chapter problems are provided throughout the book to reinforce fundamental principles.

Les Kirkup is an Adjunct Professor in the Faculty of Science at the University of Technology Sydney, and Honorary Professor in the School of Physics at the University of Sydney. He has devised laboratory programmes and taught extensively in undergraduate laboratories. He has been recognised for his work in supporting student learning in laboratories with two national fellowships and the Australian Institute of Physics Education Medal.

"The first edition of Les Kirkup's classic was for years the only set book for our first-year engineering students. This new edition reflects the rapid progress in digital instrumentation, data processing and communications since then. Even for a generation used to the minimalist style of online help, Kirkup's style makes it an easy read as well as an essential reference for error assessment. Its greatest strength lies in helping the reader to foresee experimental options which avoid the pitfalls of experimental error *before* they require correction. With so many examples, drawn from so many fields, every STEM subject student needs this book!"

— Pat Leevers, *Imperial College London*

"*Experimental Methods for Science and Engineering Students* is the ideal textbook to give students a solid foundation in the difficult tasks of designing, conducting and interpreting the results of experiments. I have used it as the textbook on my undergraduate course in Experimental Methods for several years. The students found it to be an excellent resource filled with useful and approachable examples. The updated edition has expanded to aid students in report writing and making presentations about their results. The new edition also introduces modern hardware used for data acquisition and covers data analysis in common software packages. This book is a comprehensive guide helping students develop all the skills necessary to become a competent experimentalist."

— Dr. John Kennedy, *Trinity College Dublin*

"Having used and appreciated the first edition of Les Kirkup's book, it's great to see this update. It covers the whole process of experimental work. Highlights are how the book explains the fundamentals of 'how it works' in relation to: the role of technology in modern experimentation, dealing with uncertainties in data, and a range of science communication formats that are commonly used by students and professionals in science and engineering. Techniques are helpfully illustrated with worked examples, using data from experiments that are relevant to the target students. Exercises, with answers, provide a valuable resource for practice. Les Kirkup has a clear writing style, and his extensive experience in teaching laboratory science shows in the advice that he provides about situations where students often feel confused about what to do. This book encourages students to think about their experiments, in a way that enables them to improve their outcomes."

— Margaret Wegener, *The University of Queensland*

Experimental Methods for Science and Engineering Students

An Introduction to the Analysis and Presentation of Data

SECOND EDITION

Les Kirkup

University of Technology Sydney

CAMBRIDGE
UNIVERSITY PRESS

CAMBRIDGE
UNIVERSITY PRESS

Shaftesbury Road, Cambridge CB2 8EA, United Kingdom

One Liberty Plaza, 20th Floor, New York, NY 10006, USA

477 Williamstown Road, Port Melbourne, VIC 3207, Australia

314–321, 3rd Floor, Plot 3, Splendor Forum, Jasola District Centre, New Delhi – 110025, India

103 Penang Road, #05–06/07, Visioncrest Commercial, Singapore 238467

Cambridge University Press is part of Cambridge University Press & Assessment,
a department of the University of Cambridge.

We share the University's mission to contribute to society through the pursuit of
education, learning and research at the highest international levels of excellence.

www.cambridge.org
Information on this title: www.cambridge.org/9781108418461

DOI:10.1017/9781108290104

The first edition of this book was published by John Wiley & Sons, Inc. in 1995
Second edition published by Cambridge University Press & Assessment 2019

A catalogue record for this publication is available from the British Library

Library of Congress Cataloging-in-Publication data
Names: Kirkup, Les, author.
Title: Experimental methods for science and engineering students : an introduction to the
    analysis and presentation of data / Les Kirkup (University of Technology Sydney).
Description: Second edition. | Cambridge, United Kingdom ; New York, NY : Cambridge
    University Press, 2019. | Includes bibliographical references and index.
Identifiers: LCCN 2019010796 | ISBN 9781108418461 (hardback ; alk. paper) |
    ISBN 1108418465 (hardback ; alk. paper)
Subjects: LCSH: Physical sciences–Experiments–Data processing. | Engineering–Experiments–
    Data processing.
Classification: LCC Q182.3 .K57 2019 | DDC 530.072/4–dc20
LC record available at https://lccn.loc.gov/2019010796

ISBN   978-1-108-41846-1   Hardback

To Janet

Contents

Preface to the Second Edition

I'd like to thank Simon Capelin and Cambridge University Press for the opportunity to update this book. I also thank all those who commented on the first edition and encouraged me to review and revise the text.

My goals in revising the text were to build on the methods appearing in the first edition that remain the backbone of experimental work, while including other topics that have assumed greater prominence because of, for example, changing emphases within science and engineering courses. I have taken the opportunity to include recent references and suggestions for further reading, where these expand consideration of the topics in this book. There is an emphasis on examples and exercises that reinforce principles, as those principles are introduced.

Like its predecessor, this edition seeks to assist students in their experimental work, especially students in their early years of study in a college or university course in science or engineering who are required to engage in laboratory (or field-based) work. That work might include (a) carrying out experiments where the aims of the experiments and the methods are prescribed, or (b) open-ended experiments, which give students the scope and responsibility to decide what they will do and how they will do it. As with the first edition, several topics in the book, such as reporting findings of experiments, should be of relevance to those who carry out experiments up to senior stages of an undergraduate course.

The new edition recognises technological advances that have occurred since the first edition was published with respect to carrying out experiments, and the increased emphasis on reporting findings using various modes of communication. I have indicated where useful information relating to experimental methods can be found on the Internet. I recognise that for all its virtues, the Internet can be frustrating when material is moved or removed. I apologise if or when any links I have included become 'broken'.

The chapter on the pocket calculator in the first edition has been removed, as the type of calculator featured is no longer available. With the enormous range of calculator options available to students, including calculator apps for smartphones, it is difficult to write a chapter on calculator use that would be of benefit to a large fraction of readers.

The students I have worked with have stimulated me to think carefully about the resources, examples and advice that will be useful to them as they plan, carry out and report experiments. More broadly, I hope the book will encourage a heightened awareness of the power, utility and transferability of the abilities students acquire through doing experimental work. Such abilities are of enduring value, irrespective of a student's career destination.

I have enjoyed writing this book enormously. I have sought to harness what I have learned from my students, colleagues, the writings of other authors and my own experiences of laboratory work to create a resource sympathetic to the needs of science and engineering students.

I would like to acknowledge those who have influenced this edition, offered encouragement and suggestions for inclusion or provided ideas for exercises and problems:

Mike Cortie, Yvonne Davilla, Angus Gentle, Neela Griffiths, Shirin-Rose King, Andrea Leigh, Juliette Massicot, Blair Nield, Lauren O'Brien, Anoj Poudyal, Greg Skilbeck (University of Technology Sydney)

Julie-Ann Robson, Pauline Ross (University of Sydney)

Natalie Williamson (Adelaide University)

Kate Wilson-Goosens (Australian Defence Force Academy)

Anna Wilson (University of Stirling)

Maria Parappilly, Karen Burke da Silva (Flinders University)

Andy Buffler (University of Capetown)

Gerard Ezcurra (Vernier International)

John Cadogan (Scientrific).

I would also like to thank David Raul, Jennifer Soriano and Shital Hamal for the convivial environment in which I completed much of this book.

Les Kirkup

Preface to the First Edition

As an organiser and demonstrator in undergraduate laboratories, I have never been entirely satisfied with the ad hoc arrangements made to assist those new to experimentation so that they may come to grips with vital topics such as uncertainties, graphing and keeping a laboratory notebook. It was this dissatisfaction, coupled with a belief that other topics such as report writing and the use of computers in the gathering, presentation and analysis of experimental data should be given more emphasis in texts concerned with experimental methods, that encouraged me to put pen to paper.

Experimental work should be amongst the most stimulating and satisfying activities in any course of study in science or engineering. A well-designed and executed experiment followed by *proper data analysis* offers insights into a process or phenomenon which are unlikely to be provided by a theoretical study alone. It is what goes into the *proper analysis* of experimental data which dominates the subject matter of this book. I have tried to describe the tools and techniques that will assist those engaged in presenting and analysing experimental data as part of their undergraduate studies in science or engineering.

This book is aimed at those beginning an undergraduate course of study in the physical sciences or engineering who have a significant laboratory component to their subjects. Early chapters assume very little prior knowledge of the techniques of data analysis. However, several topics in the book go beyond first-year level, and many of the matters raised should be of relevance to those engaged in experimental work up to senior stages of an undergraduate course.

With regard to the formulae that appear in the text, I have preferred to emphasise the background and assumptions relevant to a particular formula rather than to deal in depth with the mathematics of its derivation. Wherever possible, discussion of a formula is followed up with an appropriate 'worked' example. I have given references in the text to where greater details concerning formula derivation can be found.

The book divides into three sections:

(i) Chapters 1 to 4 cover the basic groundwork appropriate to first-year studies in science and engineering in the areas of notebook keeping, characterising experimental data, graphing and uncertainties.

(ii) Chapters 5 to 7 discuss the statistical analysis of experimental data and the important topic of least-squares fitting of functions to data as well as addressing the crucial area of communicating the findings of an experiment in the form of a report. Though 'long' reports are generally given more emphasis beyond first year, I feel that it is a topic of such importance that it deserves fuller treatment than a brief mention of the headings found in a report. The material contained in these chapters is likely to be most useful to data analysis problems beyond first-year undergraduate level.

(iii) Chapters 8 and 9 deal with the application of technology to the analysis and presentation of data. Though pocket calculators have been widely available for over 20 years, their usefulness for analysing experimental data is rarely dealt with explicitly in a book of this type. Superseding the calculator in power is the microcomputer, and the availability of excellent hardware and software for presentation and analysis purposes has encouraged me to write about these matters.

Because of the diverse range of topics that appear in this book I have included, in an appendix, details of other books which can be referred to for more information. As microcomputers are now commonly found in undergraduate laboratories assisting with data gathering as well as analysis, the last appendix offers an introduction to this topic.

I have gained enormous enjoyment from writing this book (and learned a few things along the way!) and look forward to any feedback it may provoke. Suggestions from students have been mostly heeded (more examples, more examples!) and I hope this has made for a book in keeping with their needs.

There are many people who have helped make this book possible. In particular I would like to thank my publishing editor, Derelie Evely, for her enthusiastic support of the project and her deadlines. Equal thanks are owed to my university for granting me the time away from my normal duties to complete the book, and to Professor David Rawson of the University of Luton, England for providing a congenial environment in which to carry out some of the work. I'd also like to thank the following reviewers and providers of ideas for examples, experimental data and problems:

Brian McGinnes (University of Sydney)
Graham Russell (University of New South Wales)
Roger Rassool (University of Melbourne)
George Haig, Leon Firth (University of Paisley)
Barry Haggett, Bill Roe, John Plater, Rosalyn Butler (University of Luton)
Bob Cheary, Jeff Kershaw, John Bell, Maree Gosper, Nick Armstrong, Patsy Gallagher,
    Peter Logan, Ray Woolcott, Walter Kalceff (University of Technology Sydney).

I addition, I would like to express my gratitude to Janet Sutherland, Shona Rawson and Billy Ward for many helpful suggestions

Les Kirkup

1 Introduction to Experimentation

1.1 Overview: The Importance of Experiments in Science and Engineering

Scientists and engineers devote valuable time and resources to experimental investigations. Why do experiments? In the first place, scientific and technical advances rely on the support that a critical experiment, or series of experiments, can offer. New and established theories are tested through experiment. Devising and carrying out an experiment that provides a thorough test of a theory or confirmation of a discovery may be challenging, but until such a test is undertaken, and the results are confirmed independently by others, the theory or the discovery is unlikely to gain wide acceptance. Additionally, carefully performed experiments may reveal new effects that require existing explanations to be modified or perhaps abandoned completely.

The outcomes of the vast majority of experiments are usually known to only a few people. However, some experiments are so groundbreaking and influential that they attract international attention. An example of this is the detection of gravity waves by a group of scientists in 2015.[1] The existence of gravity waves, produced for example by an exploding star or colliding black holes, is a prediction that emerged from Albert Einstein's general theory of relativity. It took dedicated teams of scientists and engineers many decades to conceive and build instruments sufficiently sensitive to detect those waves. Those instruments have the potential to open a new window on the universe, allowing previously hidden cosmic events to be detected. The detection and analysis of such events hold the promise of new insights into the working of the universe.

Experiments performed as part of a college or university laboratory programme or project are unlikely to attract widespread attention. Nevertheless, they provide opportunities to acquire knowledge, skills and understanding through investigating the 'real world'. Such experiments have distinct advantages over the idealised descriptions and explanations of phenomena presented in textbooks: seeing

---

[1] See Castelvecchi and Witze (2016).

something happen has more impact than reading about it. *Making* something happen, as we do when we devise and perform an experiment, is even more memorable.

Experiments are not without their difficulties; some experimental techniques take time to master and occasionally we are confronted with data that require careful examination before we are able to draw out important features. In these circumstances a little patience and persistence go a long way.

In the process of carrying out an experiment, you may need to acquire proficiency in the operation of instruments of varying degrees of sophistication. It is not the purpose of this book to give instruction on the operation of instruments, but to offer advice of a more general and, I hope, enduring nature. Specific examples and exercises are included within the forthcoming chapters where these should assist in illustrating a data analysis technique or reinforcing an important point.

An experiment may be required to assist in addressing several questions, some of a general nature and others more specific. As examples:

- How does the energy efficiency of an organic solar cell depend on its temperature?
- At what temperature does a new ceramic material exhibit superconductivity?
- How does the flow of blood through a vein depend on the diameter of that vein?
- Has the speed of light changed since the universe was formed?

It is these types of questions that form the starting point for scientific investigations.

It is instructive to outline the stages through which a typical experiment develops. Though the stages are presented here in a particular order, in practice there can be much moving back and forth between stages as new ideas emerge, underlying principles are more thoroughly understood, more sensitive equipment becomes available and experimental skills are practised and improved.

1.2   Stages of a Typical Experiment

The Aim

This is the starting point of an experiment. What do we want to find out? The clearer and better defined the aim of the experiment, the easier it is to do the planning to achieve that aim. The aim may contain an idea or hypothesis that we want to advance or test. However, it is not unusual to begin with a specific objective in mind and while doing the experiment discover something interesting or unexpected. This is part of the excitement of experimentation. If the outcomes of all experiments were wholly predictable there would be little point in undertaking them. However, we must be aware of the risk of becoming side-tracked and failing to complete the original work.

The Plan

Once the aim has been decided, a plan is devised for achieving the aim. Decisions are made as to what equipment is required, which quantities need to be measured and how they are to be measured.

Preparation

The preparation stage involves organising the experiment. Equipment is collected and assembled. Safety issues are identified, and risks assessed and controlled for.[2] This might consist of, for example, wearing the appropriate personal protection equipment when dispensing a cryogenic liquid, or wearing safety glasses when operating a laser. If the experimental technique or instrument to be used is unfamiliar, instruction should be taken from an experienced user. This avoids wasting time and making avoidable blunders, as well as reducing the risk of damaging the equipment (and possibly yourself!).

Preliminary Experiment

Once the equipment has been assembled, a preliminary experiment is often performed. This promotes familiarity with the operation of the equipment, indicates which features work well and which need further refinement and gives a feel for what values to expect when the experiment is performed more carefully. The insight that a preliminary experiment offers can sometimes lead to reconsideration of the experimental method being used. It is fair to say that experienced scientists and engineers habitually, and sometimes obsessively, seek ways to improve their experiments.

Collecting Data

The data collection phase begins. Alertness and attention to detail at this stage tend to reward the experimenter with a valuable set of data. There is little more frustrating than spending an afternoon collecting data only to discover that some omission, such as neglecting to record the units in which the measurements were made, has rendered the data unusable.

Repeatability

The experiment is carefully repeated to verify whether the first set of data is representative and can be reproduced one or more times. The repeated experiment cannot be expected to generate *exactly* the same data. However, gross variations between sets of data is a warning that further investigation is warranted.

---

[2] Colleges and universities often devise their own laboratory safety guideline documents. An example of general laboratory safety guidelines can be found at https://www.osha.gov/Publications/laboratory/OSHA3404laboratory-safety-guidance.pdf

Analysis of Data

When data collection is complete, a most important question is asked: 'what do the data tell me?' If an experiment was performed with some hypothesis in mind, then this will indicate what data analysis method(s) should be adopted. For example, in an experiment to study the change in light intensity as the distance from the light source increases, a prevailing theory may lead us to believe that there is a power law relationship between intensity, I, and distance, d, where d is the distance between the light source and the detector. The power law relationship may be written

$$I = Ad^n,\qquad\qquad(1.1)$$

where A and n are constants. The experiment would consist of measuring the light intensity as the distance between the light source and detector is varied. From the analysis of the data we would like to know how well equation 1.1 describes the relationship between I and d.

What Do the Data Tell You?

Once the data have been gathered and analysed, it is time to ask whether they are consistent with the initial hypothesis or whether the evidence is inconclusive or even contradictory. For example, in the light experiment described above, do the data provide us with enough evidence to be able to conclude that equation 1.1 *is* a good description of the relationship between intensity and distance?

Reporting the Experiment

When the experiment is complete, it is time to communicate what was done and what was found, in a clear and concise way. A report may be prepared describing the important features of the experiment such as the aim, method, data gathered, analysis method, discussion and conclusion.[3]

1.3   Documenting Your Work

An experiment may take as little time as an hour or extend over a considerable period of time. During this time, procedures may be devised and revised, apparatus assembled, data gathered and other steps taken, large and small, before the experiment is complete. Irrespective of the duration or complexity of the experiment, one thing is certain: the better the record of what has been done, the easier the task of presenting the work, perhaps in the form of a report to a laboratory supervisor. A convenient way of documenting work is to use a laboratory notebook.

---

[3] Report writing is discussed in Chapter 7.

1.3.1   The Laboratory Notebook

A laboratory notebook, also sometimes referred to as a logbook, contains a permanent record of experiments performed. For many scientists and engineers the notebook is an indispensable element of experimental work. In it they record details of an experiment, whether or not those details seem important at the time. Often it is not until sometime later that it emerges which are the important entries and which are of less value.

Notebooks are available in various sizes, but those with alternate pages of lined paper followed by graph paper are convenient to use in situations in which graphs need to be plotted. An alternative is to use a standard lined notebook and fix in graph paper as necessary. A hardback notebook, though more expensive than the softback variety, is a good investment as it tends to be more durable. In addition, there is generally less tendency for people to tear out pages from a hardback notebook. Such an act is frowned upon in all circumstances.

Though it may be tempting to write notebook entries in pencil so that they can be erased in the event of a mistake, a better option is to write consistently in ink. If a mistake has been made, a simple strike-through line can indicate an entry that should be ignored. This allows the entry to be read if it turns out sometime later not to be a mistake after all.

Many laboratory notebooks are not models of neatness (mine included) but they should represent a faithful record of what you planned, did and thought about while carrying out the experiment.

An electronic notebook is an alternative to a paper-based notebook. Electronic laboratory notebooks are increasing in popularity and have several attractive features. These include easily making backup copies of the notebook, the option of sharing the notebook with other people through the Internet and facilities for quickly searching through its contents.

Whether you use a paper or electronic laboratory notebook, it needs to be intelligible to at least one person – you! However, there may be situations in which the contents of the notebook form part of an assessment of an experiment or project you have carried out. In this case you need to remember that the notebook is going to be examined by someone else, so a logical layout of the account of the experiment is recommended. The order and description of elements that make up a typical account of an experiment are given in Table 1.1.

It is useful to number the pages of the notebook and to include a contents page. Page numbers are helpful when you want to refer to another piece of work in the notebook, and are especially worthwhile when describing a series of experiments covering several pages. For example, you might want to say *'the circuit used is that shown on page 27'*.

As described in Table 1.1, the notebook is performing two functions. The first is to document relevant information concerning the experiment, and the second is to

Table 1.1 Description of notebook contents

Notebook entries	Description
Date	Including this is good housekeeping and allows you to chronologically connect the contents of your notebook to other parts of your work. For example, it might be important to associate entries in your notebook with data that you gathered and stored on a computer on the same day.
Title	Leave the reader in no doubt as to what the experiment is about.
Aim of the experiment	Though this is something that may have already been decided on your behalf, it is so important that it bears repetition and should be given a prominent place in your notebook after the title of the experiment.
Description of instruments and apparatus	For many experiments, a brief list of apparatus used is sufficient. Recording details, such as the accuracy of an instrument, as provided in a manufacturer's calibration certificate, is good practice.
Sketch of apparatus	A fully labelled diagram of the experimental arrangement is worth a page of explanation and can assist in recalling the experiment long after it has been completed. A simple line diagram drawn freehand is all that is required.
Experimental method	If the method is given in the form of a set of instructions to follow, then the instructions can be 'cut and pasted' into the notebook. It is possible that you will have to depart from prepared instructions in some way or need to add to them, perhaps for the purpose of clarification. Such clarifications or additions need to be documented in your notebook. If you devise the method yourself then include enough detail so that you, or someone else, could repeat the experiment.
Data	It is usual to present data in tables, taking care to give each column in the table a heading which includes the unit of the quantity measured. There will be some uncertainty in every measurement you make.[a] An estimate of the uncertainty can be conveniently located in the heading of the table. Measured values are recorded *directly* into the notebook as they are made. The temptation to jot the values on scraps of paper should be resisted, as these are easily lost. Record the 'raw' data you collect. For example, do not convert units 'in your head', such as voltage values from millivolts to volts before recording them.
Comments/ Reflections	Thoughts or ideas may occur to you as your experiment proceeds, for example on how the experiment might be redesigned to improve the quality of the data. Such thoughts, as recorded in your notebook, can act as a source of inspiration when you perform another experiment. Importantly, comments recorded in your notebook may stimulate insights which can be included in a report of the experiment.
Graphs	Graphs are superior to tables for giving you 'the big picture' of the data and are often the first thing that someone examining your work will look at. Graphs must be presented properly with title, labelled axes and units.[b]
Calculations	If you need to make calculations based on your data, state the equation or relationship you are using. Work through the calculation as fully as possible in your notebook, making explicit the steps you have taken.

Table 1.1 (*cont.*)	
Notebook entries	Description
Summary/ Conclusion	A laboratory notebook is not usually the place where you will present a detailed discussion of your work. However, at the end of the experiment it is standard to include a brief summary. For example, if the aim of the experiment was to find the efficiency of a silicon solar cell, you might report: *Based on the observations made in this experiment, the maximum energy efficiency of the amorphous silicon solar cell we studied was (3.1 ± 0.3)%.*

[a] We will consider uncertainty in measurement in Chapters 4 and 5.
[b] We will consider graphing in Chapter 3.

present it in a clear way so that it might be reviewed or assessed by others. While both functions are very important, as the length and complexity of experiments increase, these functions tend to become separated and the role of the notebook changes, with the emphasis remaining on making a faithful record of an experiment.

1.3.2 Example of Pages from a Notebook

If you are not familiar with laboratory notebooks, you might be curious as to what the contents of a typical notebook look like. In Figure 1.1 I offer two pages from one of my notebooks. The work relates to a study of the performance solar cells, the first stage of which was to make measurements on an amorphous silicon cell. The aim was to establish the output power and efficiency of the solar cell when illuminated by sunlight.

The pages shown in Figure 1.1 include details of the aim of the experiment, equipment used, the method and some data. It also contains a question requiring further investigation: has the increase in the temperature of the cells over the duration of the experiment affected the output power of the cells? It is quite common that, as an experiment is being performed, it inspires other questions that deserve attention.

Writing a report is so much easier when all the details you require are in one place – your laboratory notebook. It is unlikely that the notebook by itself will make complete sense to others, but it must make sense to *you* as it is from the notebook that a report which will be read by others can be prepared.

1.3.3 Documenting Open-Ended Experiments and Project Work

So long as experiments are short and reasonably self-contained, the approach to presenting the work in your notebook described in Section 1.3.1 works well. But what do you do if you are asked to devise an experiment requiring you to be responsible for the aim and method as well as gathering data, assessing uncertainties and so on? This

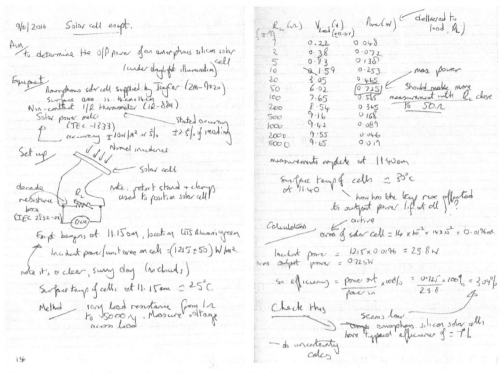

Figure 1.1 Example of pages from a laboratory notebook.

situation may well occur when doing open-ended experiments or when carrying out a project. Though the section headings appearing in the left column of Table 1.1 remain helpful, it is unlikely that you will be able to fill out the notebook in such a stepwise manner. Despite this, concisely and clearly documenting open-ended experiments and projects in a notebook is vital, as much of the information you will draw on when reporting the work to others will reside in your notebook.

1.4   Comment

In some circumstances a laboratory notebook can be unexpectedly valuable. If you are lucky enough to make a scientific discovery or technical advance, it is likely to appear first in your notebook. If, in addition, you are unlucky enough to be in competition with someone who claims to have made the discovery or advance first, you might find that your notebook becomes a vital piece of legal evidence to show what you did and when. For this reason, it is common for experimenters in industrial or government laboratories to have the contents of their notebooks verified regularly by a co-worker or supervisor. Beware, prosperity and reputation could be tied to your laboratory notebook, so use it well and keep it safe!

2 Characteristics of Experimental Data

2.1 Overview: What are the Important Features of Experimental Data?

Measurements made during an experiment generate data which are recorded, analysed and reported. We need to display numerical data in ways which assist in analysis of those data. In addition, it is desirable to be able to examine the data as a whole so that trends can be recognised, for example the existence of a linear relationship between measured quantities. A table is an effective way of presenting data requiring manipulation, while a well-drawn graph is a revealing pictorial representation.

There are several questions to consider while gathering experimental data:

(i) What is the unit associated with each measurement?
(ii) How much variability is there in the data?
(iii) What should be included in a table of data?
(iv) Can we estimate or anticipate the size of a quantity before it is measured?

We now turn our attention to these and related questions.

2.2 Units of Measurement

At the heart of an experiment lies measurement and measurement requires a system of units. When a scientist or engineer claims to have made a breakthrough, perhaps developing an alloy with excellent corrosion resistance at high temperatures, other workers in that area of science or engineering want to know as many details of the characteristics of this material as possible. These might include its melting point, density, thermal conductivity, heat capacity and crystal structure.[1]

---

[1] A quantity, such as the melting point of an alloy, that we wish to determine through measurement is sometimes referred to as the *measurand*.

Table 2.1 Fundamental SI units

Quantity	Name of unit	Symbol
Mass	kilogram	kg
Length	metre	m
Time	second	s
Electrical current	ampere	A
Temperature	kelvin	K
Luminous intensity	candela	cd
Amount of substance	mole	mol

Above all, *quantitative* estimates of the properties are needed. These permit measurements made in other laboratories (perhaps on the same alloy in another part of the world) to be compared directly with the original work. A starting point for that comparison is that everyone agrees on a set of units of measurement and uses those units consistently when the results of experiments are reported.

2.2.1 The SI System of Units

Globally, the most widely used system of units in science and engineering is the Système Internationale or SI system, and it is the one we will use throughout this book.[2] At its core are seven fundamental units, as shown in Table 2.1. Other units are derived from the fundamental units. An example of a derived unit is the metre per second (written m/s or $m\,s^{-1}$), which is a unit of velocity.[3]

The International System of Units by BIPM[4] is an excellent source of information on the fundamental units, how they are defined and how they may be combined to give a host of derived units. Many of the derived units have been given names which may be familiar to you. As examples, the newton, volt, watt and joule are all derived units in the SI system.

In pursuit of our goal of effective presentation and analysis of data, we should remember this: *Whenever we fill a table with data, plot a graph or make a remark concerning measurements or calculations based on those measurements, we must* ***always*** *state the units in which we are working.*

---

[2] We will not use SI units exclusively. Units such as the minute and the degree (which are not SI units) are so commonly used that examples will be given that use these units.

[3] Note that it is equally acceptable to write derived units using the slash mark, /, as in m/s, or using exponents, such as $m\,s^{-1}$, and is generally a matter of preference which you use. I prefer using the slash mark except where it is clearer to use exponents.

[4] BIPM stands for Bureau International des Poids et Mesures. The BIPM document detailing the International System of Units can be found at http://www.bipm.org/en/publications/si-brochure/.

This can be put another way. Tables, graphs and calculations are meaningless unless the units of the quantities are stated. A critical question we need to ask when about to use an instrument is: 'In what unit does this instrument measure something?' When using a stopwatch, the answer may be immediately apparent. When using a less familiar instrument, such as a vacuum gauge, the answer may be less obvious.

2.2.2 Multiples and Subdivisions of Units

For some measurements, the fundamental unit might be clumsy, and it would be more convenient to use a multiple or subdivision of that unit. An example can be taken from the topic of electricity. The farad (symbol F) is the SI unit of electrical capacitance. Capacitors are available that lie in the range of capacitance from 1000 F or more to 10^{-12} F or less. For example, decoupling capacitors used to protect electrical circuits are typically of the order of 10^{-6} F. It would be convenient if we could assign a name to the subdivision 10^{-6} F, and in fact there already is a name: it is the *micro*farad (symbol μF). It is easier to speak of 'five point five microfarads' than to say 'five point five times ten to the minus six farads'. As 'micro' is placed in front of the unit, it is termed a *prefix*.

There are prefixes for multiples of units extending from 10^{-24} to 10^{24}. Table 2.2 contains the prefixes corresponding to each multiplier in the range 10^{-15} to 10^{12}, which will cover most situations you are likely to encounter in science or engineering.

Incidentally, while in many situations we find it convenient to express length in centimetres, it is unusual to see the centi prefix used elsewhere. As examples, it is rare to hear a voltage expressed as '2.2 centivolts' or a change in temperature as '50 centikelvins'.

Table 2.2 SI prefixes

Power of 10	Prefix[a]	Symbol	Example
10^{-15}	femto	f	fs (femtosecond $\equiv 10^{-15}$ s)
10^{-12}	pico	p	pF (picofarad $\equiv 10^{-12}$ F)
10^{-9}	nano	n	nA (nanoampere $\equiv 10^{-9}$ A)
10^{-6}	micro	μ	μPa (micropascal $\equiv 10^{-6}$ Pa)
10^{-3}	milli	m	mJ (millijoule $\equiv 10^{-3}$ J)
10^{-2}	centi	c	cm (centimetre $\equiv 10^{-2}$ m)
10^{3}	kilo	k	kV (kilovolt $\equiv 10^{3}$ V)
10^{6}	mega	M	MW (megawatt $\equiv 10^{6}$ W)
10^{9}	giga	G	GΩ (gigaohm $\equiv 10^{9}$ Ω)
10^{12}	tera	T	THz (terahertz $\equiv 10^{12}$ Hz)

[a] Note that, with the exception of centi, all the other powers of 10 that appear here are multiples of 3.

Exercise A

Express:

(i) $2.2 \times 10^{-6}\,V$ in μV
(ii) $6.2 \times 10^{-2}\,m$ in mm
(iii) $6.52 \times 10^{4}\,J$ in kJ
(iv) $1.8 \times 10^{5}\,W$ in MW
(v) $6.7 \times 10^{-11}\,F$ in pF
(vi) $8.2 \times 10^{-8}\,Pa$ in nPa
(vii) $1.44 \times 10^{-4}\,H$ in mH
(viii) $1.67 \times 10^{5}\,g$ in kg.

2.3   Tabulation of Data

Experimental observations take forms as contrasting as recording the electrical activity of the human brain using electrodes in contact with the scalp and measuring the temperature of an oil bath using an alcohol-in-glass thermometer. Whatever the measurement technique or instrument used, carefully made observations are the foundation of good experimentation. It is important that the observations are properly recorded in a manner that will permit analysis later.

There are two types of experimental situation that commonly occur. In the first, repeated measurements of a quantity are made where there is no reason to believe that the quantity is varying. Examples of this would be timing how long it takes an object to fall through a given distance or measuring the wavelengths of light emitted from a lamp containing helium gas. In the other situation we are trying to establish the relationship between two quantities, let's call them A and B. We could do this by changing A and observing the effect that this has on B. An example of this would be measuring the viscosity of a liquid (quantity B) as the temperature of the liquid (quantity A) changes.

In both situations a convenient way in which to present data, whether it be in a laboratory notebook or a report of the experiment, is in the form of a table. Table 2.3 contains the values of 10 repeated measurements of the time it took a small object to fall through a distance of 25 m.

Table 2.3 Hand timing of an object falling 25 m

Time of fall (s)	2.2	2.0	2.6	1.9	2.1	2.4	2.2	2.3	2.3	2.0

Table 2.4 Hand timing of an object falling 25 m									
Time of fall /s	2.2	2.0	2.6	1.9	2.1	2.4	2.2	2.3	2.3	2.0

For the table to be useful, it needs a distinct heading with a clearly stated unit. The unit is shown in brackets after the name of the quantity being measured. In many circumstances it is also useful to estimate the uncertainty in each measurement as it is made (to be discussed in Section 2.4) and to include that information in the table.

An alternative method of indicating the unit of measurement in a table is shown in Table 2.4. The argument for using /s instead of (s) is as follows: tables contain numbers. In this experiment we have measured time in seconds, so in order to be able to display those times as numbers in the table we must 'cancel out' the unit by *dividing* our values by seconds, hence the /s in the heading. This approach to indicating units, though favoured by many authors, can sometimes be confusing and I prefer the method of presenting units within brackets in the heading of the table.

As an experiment proceeds, values obtained through measurement are recorded in a table with no attempt at data manipulation, such as squaring values or subtracting a constant. If you *do* manipulate the data as you go along, and in the process make a mistake, it can be difficult to work back to find the origin of the mistake. A classic example of this is when converting from one form of unit to another 'in your head', say cm^2 to m^2, and wondering later why the final answer is 10 000 times too large.

2.3.1   Tabulating Data Represented in Scientific Notation

Writing the length 1.1 μm as 1.1×10^{-6} m is to express the value in *scientific notation*[5] (see Section 2.5.3 for more details). If we have data expressed in scientific notation which we wish to enter into a table, the readability of the table is improved by indicating the multiplying power of 10 in the heading of the appropriate row or column of the table.

For example, as part of an experiment, the electrical inductance of a coil of wire was measured. The following data were obtained:

$$9.5 \times 10^{-3} H, 9.3 \times 10^{-3} H, 9.9 \times 10^{-3} H, 9.9 \times 10^{-3} H, 9.1 \times 10^{-3} H.$$

Table 2.5 displays the data by showing the multiplying factor, expressed as a power of 10 in the heading to the table.

---

[5] Sometimes this is referred to as 'powers of 10' notation.

Table 2.5 Values of inductance of a coil of wire

Inductance ($\times 10^{-3}$ H)	9.5	9.3	9.9	9.9	9.1

Exercise B

Present the following pressure data in a table in the manner described in this section:

1.03×10^5 Pa, 1.01×10^5 Pa, 1.01×10^5 Pa, 9.9×10^4 Pa, 1.05×10^5 Pa, 1.08×10^5 Pa.

2.4   Uncertainties in Values Obtained through Measurement

Despite our best efforts or the quality of the equipment used, there is going to be an amount of variability in quantities measured in an experiment. It could be that during a measurement of temperature of a liquid using an alcohol-in-glass thermometer, the position of the top of the alcohol column fluctuates slightly. We may have difficulty deciding whether the alcohol is at the 15.5 °C mark or has reached the 16.0 °C mark. We shouldn't feel frustrated that we cannot make an 'exact' measurement. Exact measurement doesn't exist. We might need to look at the design of the experiment to try to ensure that we have done all we can to measure the important quantities as accurately as possible. Nevertheless, some *uncertainty* in the measured quantity remains no matter how good the experimenter or experimental design (experienced experimenters constantly seek ways to reduce the uncertainty in the measurements they make).

Suppose the fluctuating position of the top of the alcohol column in the thermometer means that no individual temperature measurement can be made to better than 0.5 °C. We can write that the temperature of the liquid is

$$(15.5 \pm 0.5)°\text{C}.$$

The $\pm$ sign indicates we believe that the temperature lies in the interval (15.5 − 0.5) °C to (15.5 + 0.5) °C; that is, somewhere between 15.0 °C and 16.0 °C.

Estimates of the uncertainty in values obtained through measurement should accompany the values and need to be recorded in your laboratory notebook. Often it is convenient to write the uncertainty, especially if it is the same for all values, in the heading of the column in the table containing the observations.

Table 2.6 shows an example of this taken from an experiment in which the electrical resistance of a copper wire was measured as the temperature increased.

Table 2.6 Variation of electrical resistance with temperature of a copper wire

Temperature (°C) ± 0.5 °C	Electrical resistance of length of copper wire (Ω) ± 0.001 Ω
8.0	0.208
16.5	0.213
23.5	0.222
32.0	0.229
40.5	0.232
54.5	0.243

There is more to be said about measurement uncertainties. If we have to combine values in order to calculate another quantity,[6] how do uncertainties in the individual values add together to give an uncertainty in the calculated quantity? This and related matters are dealt with in Chapter 4.

2.5 Significant Figures

If a value obtained from a particular experiment involving the timing of an event is recorded as 6.12 s, this implies that the actual value lies between 6.11 s and 6.13 s. If the value is written as 6.124 s, then this implies that the actual value lies between 6.123 s and 6.125 s. Writing the value as 6.12 s is to give it to three *significant figures*, and to write the value as 6.124 s is to give it to four significant figures. After making a measurement, the first inclination is to record a value which contains as many figures as the instrument provides. We should, wherever possible, assess and record the uncertainty in the measured value as discussed in Section 2.4 (and more fully in Chapter 4). Omitting this information obliges the reader to make an educated guess as to the size of the uncertainty based on the number of significant figures presented.

If the number 0.0010306 is recorded during an experiment, how many significant figures are implied by the way the number is written? Count the number of figures between the first non-zero figure and the last figure inclusive. In the case of the value 0.0010306, the first non-zero figure is a '1'. There are four remaining figures, making *five* significant figures in all.[7]

---

[6] For example, we may need to calculate the density of an object by bringing together the mass of the object and its volume. How does the uncertainty in the mass combine with the uncertainty in the volume, to give the uncertainty in the density of the object? We consider such situations in Chapter 4.

[7] The zeros that precede the '1' serve to place it in the correct position with respect to the decimal point, and are not regarded as significant figures.

This rule for counting significant figures is fine except for situations in which we have whole numbers greater than 10 that end with one or more zeros, for example 20, 1670, 3400, 545 000. Unless we are told explicitly that for the number 1670 all of the figures are significant, it is usual to infer that the number is known only to the nearest 10, and therefore the zero is *not* significant. It follows that 1670 is a number known to three significant figures. This will be revisited in Section 2.5.3.

Example 1

How many significant figures appear in the following numbers?
(i) 1.654, (ii) 0.00437, (iii) 64 000, (iv) 1.20, (v) 0.10000738, (vi) 0.010, (vii) 50.0010, (viii) 75.1

Answers

- (i)   four
- (ii) three: the first non-zero figure (in this case the '4') is the first and most significant figure
- (iii) two
- (iv) three: if the zero is not significant then the number should have been written as 1.2
- (v)   eight
- (vi) two
- (vii) six
- (viii) three

Exercise C

How many significant figures are implied by the way the following numbers are written?

- (i)   3.24
- (ii)   0.0023
- (iii) 83 400
- (iv)   1.010
- (v)   10.5
- (vi)   900.10
- (vii) 43.350
- (viii) 0.00100
- (ix)   1.00

2.5.1 Rounding Numbers

When the result of a calculation produces a number that has many figures (which can easily happen if you use a scientific calculator), we may need to reduce the number of figures that appear. If, for example, the number 1.3563342 must be reduced to two significant figures (we term this 'rounding' the number), a decision must be made as to whether the second figure (in this case the '3') should be left as it is or increased by 1. This is done by considering the third figure. If that figure is 5 or greater, the second figure is rounded up. If the third number is less than 5, the second figure is left alone. So, in this example, the number 1.3563342 to two significant figures would be 1.4.

In a calculation involving several arithmetic steps, it is very good advice *not* to round numbers until all the calculations have been completed; otherwise the rounding process itself can have a large effect on the numbers that emerge from the calculations (an example of this is given in Section 6.2.2).

Exercise D

Round the following numbers to three significant figures:

   (i)   18.92
  (ii)   0.10759
 (iii)   725.4
 (iv)   1.7602
  (v)   62 654
 (vi)   0.0072491
(vii)   42 045
(viii)   1055

2.5.2 Calculations and Significant Figures

Suppose you are required to find the circular cross-sectional area of a cylindrical rod after measuring its diameter to be 8.9 mm. The relationship between the area of a circle, A, and its diameter, d, is

$$A = \frac{\pi d^2}{4},$$

so that[8]

$$A = \frac{\pi (8.9 \text{ mm})^2}{4} = 62.21138852 \text{ mm}^2.$$

---

[8] When calculations involve quantities with units, we will include the units in any arithmetic steps shown.

The calculation was done using a calculator capable of giving answers with up to 10 figures. There is something disturbing here; d is known to two significant figures, and yet the area is given to 10 significant figures. Is this reasonable? The answer is no. If we had a computer that gave results of calculations up to 100 figures long we could have written A to that many figures, but what would they mean? The answer is: absolutely nothing!

If you are required, as in the example above, to perform a calculation in which the uncertainties in the quantities are not known, and all you have to work with is the number of significant figures in each quantity in the calculation, then the following rules are useful:

Rule 1: when multiplying or dividing numbers, *identify the number in a calculation that is given to the least number of significant figures. Give the result of the calculation to the same number of significant figures.*

For example, multiplying 3.7 by 3.01 gives 11.137. The number 3.7 has the least number of significant figures (two) and so we should give the answer as 11.

Rule 2: when adding or subtracting numbers, *round the result of the calculation to the same number of decimal places as the number in the calculation given to the least number of decimal places.*

For example, adding 11.24 and 13.1 gives 24.34. Using Rule 2 we must give the result to one decimal place, that is, as 24.3.

Exercise E

Using the rules given in this section, write down the results of the following calculations to an appropriate number of significant figures:
  (i) 1.2×8
  (ii) 13.0×43.23
  (iii) 0.0104×0.023
  (iv) $33 + 435.5$
  (v) $14.1/76.3$
  (vi) $105.55 - 34.2$
  (vii) $32.0/4.95$
  (viii) $12 + 87.5 + 1.759$
  (ix) $-196 + 25.2$

2.5.3   Significant Figures and Scientific Notation

It is not always clear how many figures in a number are significant. By changing the unit in which a number is expressed, it can appear that the number of significant figures changes. For example, suppose in an experiment a time interval was recorded as 346 s. We could choose to write the time in other units such as milliseconds or microseconds. These would be written as 346 000 ms and 346 000 000 µs, respectively. In both cases the number of significant figures remains as three. However, if someone asked you for your value for the time interval to be expressed in ms, how would they know that, in your value of 346 000 ms, only the first three figures were significant? It is possible that you used a timing device capable of a resolution of 1 ms and that the time interval came out, to the nearest millisecond, as 346 000 ms; that is, all six figures are significant. The way to get around this difficulty is to present numbers in *scientific notation.*

In scientific notation the first non-zero figure that appears is followed by a decimal point, so that 346 becomes 3.46. To bring the number back to its original value we must multiply 3.46 by 100 or 10^2. We can now write the time interval as 3.46×10^2 s. In terms of milliseconds and microseconds this becomes 3.46×10^5 ms and 3.46×10^8 µs, respectively.[9] The number of significant figures is equal to the number of figures that appear to the left of the multiplication sign.

In situations where a number lies between 1 and 10, for example 7.15, we could write this as 7.15×10^0. Though this is technically correct, it is far more usual for the number to be expressed simply as 7.15.

Table 2.7 contains a variety of numbers and their representation in scientific notation (we assume here that all the figures given are significant).

Exercise F

(1) Give the following numbers to *four* significant figures in scientific notation:
    (i) 0.0056542
    (ii) 125.04
    (iii) 93 842 773
    (iv) 3 400 042
    (v) 0.000000100092
    (vi) 0.57236
    (vii) 6376.7
    (viii) 0.0095435
(2) Repeat part (1), but this time give the numbers to *two* significant figures.

---

[9] A large number such 3.46×10^5 may be written as 3.46E5 (which means the same thing), or a small number such as 1.3×10^{-8} may be written as 1.3E−8. Computers often display very large and very small numbers this way.

Table 2.7 Examples of numbers expressed in scientific notation

Number	In scientific notation
12.65	1.265×10^1
0.00023	2.3×10^{-4}
342.5	3.425×10^2
34 001	3.4001×10^4

2.6   Estimation

We often have a feel for the size of values that should emerge from an experiment, before it is carried out. This may come from having performed a similar experiment in the past, or just from everyday familiarity with the quantity being studied. As a result, it is natural to be suspicious when the values we obtain through experiment differ considerably from our expectations.

For example, if we were to measure the velocity of a car moving along a main street and found it to be 4×10^6 m/s, we should suspect something to be wrong. Equally, when measuring the mass of a small glass beaker we should look again if that mass was recorded as 45.64 kg.

The habit of estimating the expected value to within an order of magnitude is helpful for avoiding gross (and sometimes embarrassing) mistakes. Imagine you are given a battery that has been removed from a LED torch and are asked to determine the output voltage of the battery using a voltmeter. People will be amused if you report that voltage as 1500 V, and may not take anything else you say seriously.

2.6.1   Fermi Problems

Estimation is also valuable in situations where:

- A quantity cannot be measured without a great deal of time, effort or specialised equipment.
- A rough value of a quantity is required before an experiment is performed (allowing for comparison of the value once the experiment has been performed). This assists in identifying and correcting mistakes.
- The determination of the value of a quantity requires 'intelligent guesswork'. This might happen, for example, where approximations must be made before a calculation of the value is possible.
- Access to detailed information is limited that would allow for the calculation of a quantity, requiring you to rely on knowledge 'in your head' combined with an understanding of scientific or engineering principles.

In many cases, estimating a quantity within a factor of 2 or 3 of the quantity which emerges from a more extensive analysis (where all the approximations and assumptions had been thoroughly examined and fine-tuned, if necessary) is regarded as a success.

Fermi problems, which generally require a combination of estimation and physical insight, are named after Enrico Fermi. Fermi was a Nobel-prize winning physicist, active in the twentieth century,[10] famous for posing difficult scientific problems and solving them through a series of assumptions and approximations. In contrast to textbook problems, which are usually well defined and lead to a unique answer, Fermi problems generally require an amount of interpretation. They invite a person to guess, approximate and justify the approach they adopt, as well as to defend the answer that emerges for the value of a quantity. In many ways this process describes an aspect of what scientists and engineers do for a living.

Example 2
How much energy is required to raise the temperature of the water in an Olympic-sized swimming pool by 5 °C?

Answer
Principle
We begin by using the relationship between energy, Q, transferred to a body (in this case to the water in a swimming pool) and the temperature rise, θ, caused by the heat transfer. That relationship is $Q = mc\theta$, where c is the specific heat of the water and m is the mass of the water in the pool.[11]

Values to be Input to the Calculation
A rough value for the specific heat of water is 4000 J/(kg °C). Note that in a Fermi problem, you only need to know (or guess) values to one significant figure.

One approach to estimating the mass of water in the pool is to:

(1) estimate the pool's dimensions in order to find its volume, V
(2) find the mass of water using $m = \rho V$, where ρ is the density of the water.

An Olympic-sized pool is 50 m long. Assuming the pool has nine lanes, each of width 2 m, then the width of the pool is about 18 m. Let's call it 20 m to one significant figure. I estimate that the pool has a depth of more than 1 m, but less than 5 m. Let's call it 3 m.

---

[10] An introduction to Fermi problems can be found at http://en.wikipedia.org/wiki/Fermi_problem
[11] See Walker (2014), chapter 18, for an explanation of $Q = mc\theta$.

Estimate of volume = 50 m × 20 m × 3 m = 3000 m^3.
The density of water is close to 1 g/cm^3 = 1000 kg/m^3.
So, the mass of water is $m = \rho V$ = 1000 kg/m^3 × 3000 m^3 = 3 × 10^6 kg.
Bring it all together using $Q = mc\theta$ = 3 × 10^6 kg × 4000 J/(kg °C) × 5 °C = 6 × 10^{10} J.

Exercise G

Referring to Example 2, how long would it take to raise the temperature of the water in the Olympic-sized swimming pool by 5 °C if the heat were supplied to the water by the heating element of a domestic electric kettle? What assumptions or approximations have you made in order to carry out your calculation?

2.7   Comment

Data provide the foundation upon which understanding in science and engineering is built. A basic requirement is that data be expressed in units that are recognised and accepted internationally. In this chapter we have discussed the most commonly adopted system of units: The International System of Units. We have also considered how to express data, for example using scientific notation.

In some situations, it is possible to estimate the size of values likely to emerge from an experiment, prompting our attention to be alerted when the values obtained differ considerably from our estimate. Exploring the reasons for a discrepancy can lead to improved insight into the experiment or perhaps hint that a mistake has occurred, for example when converting units.

No matter how faithfully we record our experimental values, taking due regard of such things as uncertainties and orders of magnitude, we still face a difficulty: absorbing all the data at once so that we can identify relationships between measured quantities is not easy if the data remain in tabular form. An effective way of presenting the data to reveal relationships and anomalies is to plot a graph. We deal with graph plotting in the next chapter.

Problems

2.1. The mass of a small glass beaker was measured with a balance and found to be 45.64 g. When a liquid was added to the beaker the balance indicated 92.5 g. Give the mass of the liquid in the beaker to the appropriate number of significant figures.

2.2. In a heat transfer experiment, an amount of heat, Q, is transferred to silver at its melting point, causing mass, m, of the silver to melt. The relationship between m and Q is $Q = mL$, where L represents the heat of fusion of the silver.

Given that $L = 88 \times 10^3$ J/kg and $Q = 4550$ J, calculate the mass of silver melted in kg. Express your answer in scientific notation to the appropriate number of significant figures.

2.3. As part of an experiment, a student was required to find the density, ρ, of a small solid metal sphere. ρ is given by $\rho = m/V$, where m is the mass of the sphere and V is its volume ($V = 4\pi r^3/3$, where r is the radius of the sphere). The student's notebook entry for this part of the experiment was as follows:

I measured the mass of the sphere and its diameter and obtained the following:

Mass of sphere = 0.44 g

Diameter of sphere = 4.76 mm

Using formula for volume of sphere,

$$V = \frac{4\pi(4.76)^3}{3} = 451.761761 \ \mathrm{mm}^2$$

Therefore, density of sphere,

$$\rho = \frac{0.44}{451.761761} = 9.739647 \times 10^{-3}$$

The student's notebook entry contains a number of mistakes or omissions. Identify the mistakes or omissions and, where possible, correct them.

In answering the following Fermi problems, include any assumptions or approximations you make.

2.4. Estimate the volume of your body. Express the volume to one significant figure in

(i) mm^3

(ii) cm^3

(iii) m^3

2.5. Estimate the surface area of your body. Express the area to one significant figure in

(i) mm^2

(ii) cm^2

(iii) m^2

3 Graphical Presentation of Data

3.1 Overview: The Importance of Graphs

Our ability to absorb and process information when it is presented in the form of a picture is so good that it is natural to exploit this talent when analysing data obtained from an experiment. When data are presented pictorially, trends or features in the data can be detected that we would be unlikely to recognise if the data were given only in tabular form. This is especially true in situations where a set of data consists of hundreds or thousands of values, which is a common occurrence when a computer is used to assist in data gathering. Additionally, a pictorial representation of data in the form of a graph is an excellent way to summarise many of the important features of an experiment. A graph can indicate:

(i) the quantities being studied
(ii) the range of values obtained through measurement
(iii) the uncertainty in each value
(iv) the existence or absence of a trend in the data gathered (for example, plotted points may lie in a straight line or a curve, or may appear to be scattered randomly across the graph)
(v) which plotted points do *not* follow the general trend exhibited by most of the data.

x–y graphs (also known as scatter plots or *Cartesian coordinate* graphs) are used extensively in science and engineering to present experimental data, and it is those that we will concentrate on in this chapter.

3.2 Plotting Graphs

An x–y graph possesses horizontal and vertical axes, referred to as the x and y axes, respectively. Each point plotted on the graph is specified by a pair of numbers termed the *coordinates* of the point. For example, point A in Figure 3.1 has the coordinates $x = 20$, $y = 50$. The coordinates of the point may be written in shorthand as (x,y), which in the case of point A on Figure 3.1 would be (20,50). To assist in the

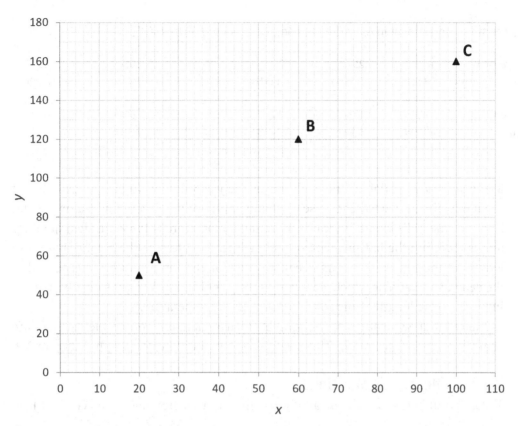

Figure 3.1 Example of an *x*–*y* graph.

accurate plotting of data, graph paper may be used on which are drawn evenly spaced vertical and horizontal gridlines as shown in Figure 3.1.

The x coordinate is sometimes referred to as the *abscissa* (or more loosely, but more commonly, the 'x value' of a data point) and the y coordinate as the *ordinate* (or the 'y value' of a data point).

To be useful for displaying experimental data, a graph requires more informative axis labels than those shown in Figure 3.1. Labels need to be attached to the axes to indicate the physical quantities being studied. Also required on the axes are the units in which the measurements were made. These and related matters are dealt with in this chapter.

3.2.1 Dependent and Independent Variables

The quantity which is controlled or deliberately varied during an experiment is referred to as the *independent* variable[1] and is plotted as the x coordinate. The quantity that varies in response to changes in the independent variable is referred to

---

[1] The quantity that is controlled or deliberately varied in an experiment is sometimes referred to as the *predictor* variable.

Table 3.1 Length of an aluminium rod in the temperature range 0 °C to 250 °C

Temperature (°C)	0	25	50	75	100	125
Length (m)	1.1155	1.1164	1.1170	1.1172	1.1180	1.1190
Temperature (°C)	150	175	200	225	250	
Length (m)	1.1199	1.1210	1.1213	1.1223	1.1223	

as the *dependent* variable[2] and is plotted as the y coordinate. To take an example: if in an experiment we were to raise the temperature of an aluminium rod, we would find that the rod's length increases. Here the temperature is the controlled quantity and is therefore the independent variable which is plotted as the x coordinate. The length of the rod increases *as a consequence* of the temperature increase and is the dependent variable which is plotted as the y coordinate. We say formally that the increase in the length of the rod is a *function* of temperature.

Table 3.1 shows values of the length of an aluminium rod as the temperature changes from 0 °C to 250 °C. These data are plotted on the x–y graph in Figure 3.2.

3.2.2   Title, Labels and Units

The graph in Figure 3.2 is typical of those found in science and engineering. It has:

(i) a title indicating the relationship being investigated. When it is stated that quantity 'A' (in this case the length) is plotted *versus* or *against* quantity 'B' (in this case the temperature), then quantity 'A' is plotted on the y axis quantity 'B' on the x axis

(ii) axes clearly labelled with the names of the quantities under study and their units of measurement. Each unit of measurement is given in brackets after the name of the quantity represented on that axis.[3]

Occasionally you will encounter graphs in which the scale on the y axis is proportional to the magnitude of the dependent quantity, but the absolute value of the quantity (that is, the quantity expressed in a recognised system of units) is not given. For example, in an experiment in which light intensity is measured as a function of position or time, we may need to know only the *relative* change in intensity as the experiment proceeds. So long as the output of the light detector used in the experiment is proportional to the intensity of light, it is not necessary to convert that output to the SI unit for intensity (which, for luminous intensity, would

---

[2] The quantity that varies as a result of changes in the independent variable is sometimes referred to as the *response* variable.

[3] If the graph appears as a figure in a report, then it is usual for a caption describing the graph to be placed adjacent to the figure.

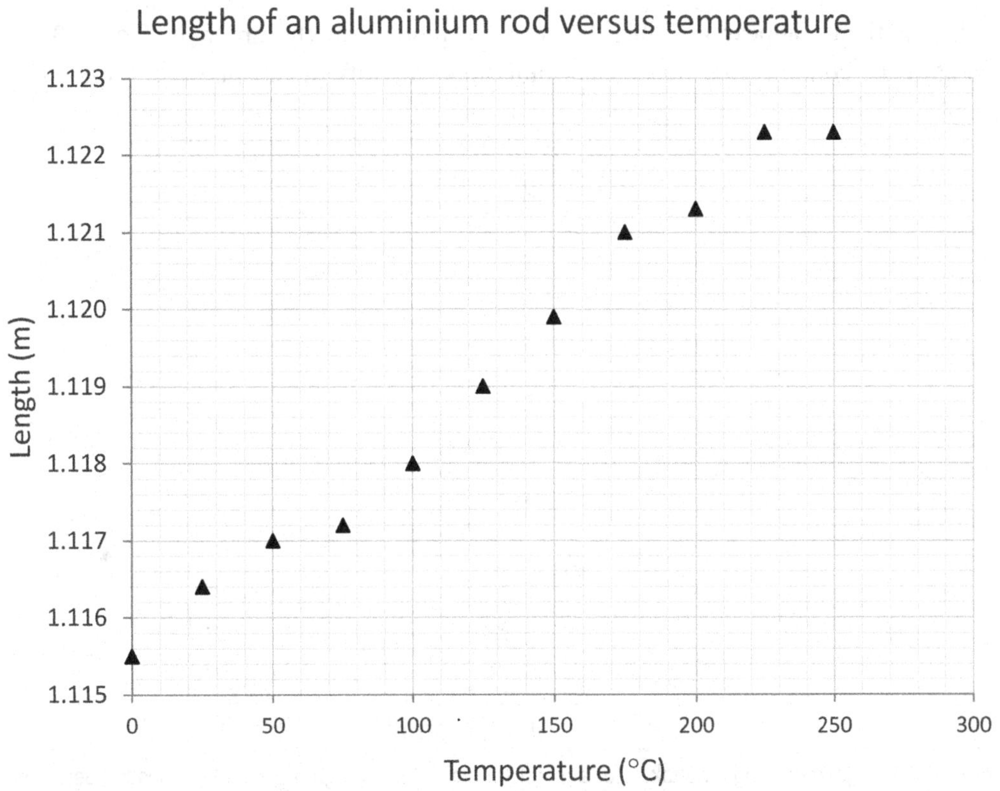

Length of an aluminium rod versus temperature

Figure 3.2 Graph of the variation of the length of an aluminium rod with temperature.

be the candela). In such a case, the y axis of the graph would be labelled *Light intensity (arbitrary units)*.

Exercise A

An experiment is performed in which the time for a small metal sphere to fall a fixed distance through a liquid is recorded as the temperature of the liquid increases. The

Table 3.2 Time taken for a metal sphere to fall through a liquid at various temperatures

Temperature (°C)	Time (s)
21	62
26	48
30	35
37	26
42	22
46	19
51	17

data gathered appear in Table 3.2 and are plotted on the graph in Figure 3.3. Figure 3.3 contains four mistakes or omissions. Identify the mistakes or omissions.

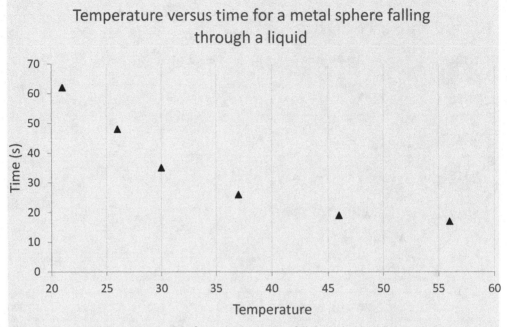

Figure 3.3 Graph indicating time taken for a metal sphere to fall through a liquid at various temperatures (with mistakes and omissions).

3.2.3 Scales, Symbols and Keys

In most situations, it is sensible to choose scales so that the plotted points occupy most of the available graph paper, leaving enough space to add labels, units and a title. The scales should be chosen to reduce the effort required to plot points,[4] and so that values can be easily read from the graph.

When indicating data points on a graph, it is preferable to make the symbol representing the data point too big rather than too small. Figure 3.4 shows three sets of data, plotted on the same axes, representing the motion of three cyclists. The points indicated by the symbols ⊙ and ▫ are easy to see, but those points indicated by small dots could go unnoticed as they are much less visible.

A key, sometimes referred to as a legend, is included on the graph when several sets of data need to be identified. The graph is generally less cluttered if the key is placed outside the gridline rectangle, as in Figure 3.4.

---

[4] Scale divisions of 1, 2 or 5 (or these multiplied by some power of 10) reduce the effort required to plot points on a graph (assuming plotting is done by hand).

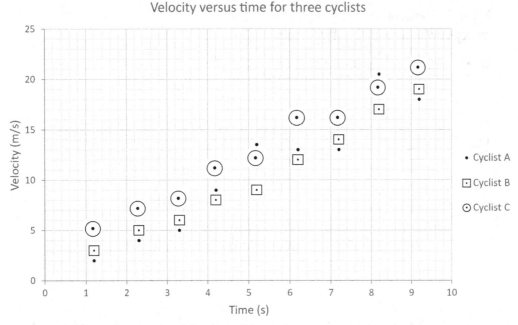

Figure 3.4 Velocity versus time graph for three cyclists.

3.2.4   The Origin

For many graphs, the numbering of both axes begins at zero, so that the bottom left-hand corner of the graph has coordinates (0,0). This point is referred to as the *origin* of the graph. There is no rule to say that we must include the origin on a graph. To do so may cause important information to be concealed. As an example, Figure 3.5 shows a graph of the data from Table 3.1 when the origin is included. The points in Figure 3.5 appear to lie almost horizontally and might lead us to conclude (incorrectly) that the length of the aluminium rod does not change with temperature. Including the origin has required the use of a y scale that is much too coarse to reveal whether the length of the rod varies with temperature. Figure 3.2, which contains the same data but does not include the origin, indicates much more clearly the relationship between length and temperature.

3.2.5   Error Bars and Line Drawing

Uncertainties in values obtained through measurement were introduced in Section 2.4. It is possible to indicate the size of the uncertainties in the x and y quantities by attaching *error bars* to each data point on the graph.[5] Error bars are vertical and/or horizontal lines that extend from a data point. The length of each

---

[5] We could call them 'uncertainty bars' instead of error bars, but the latter term is much more commonly used.

Table 3.3 Variation of temperature of an object with time as the object cools	
Time (s) ±0.5 s	Temperature (°C) ±5 °C
1	125
7	108
13	99
19	90
26	82
32	76
37	72

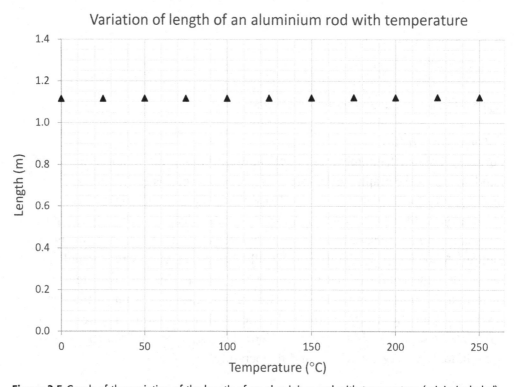

Figure 3.5 Graph of the variation of the length of an aluminium rod with temperature (origin included).

line is a measure of the size of the uncertainty in the quantity. Let us illustrate this by considering a specific example.

Table 3.3 shows data from an experiment in which the temperature of an object has been recorded as it cooled. In the heading of each column there is an estimate of the uncertainty in the measured quantity. In this example, time is plotted as the x quantity and temperature as the y quantity.

To indicate the uncertainty in the time, we draw horizontal error bars as shown in Figure 3.6a, and to indicate the uncertainty in the temperature, we draw vertical error bars as shown in Figure 3.6b.

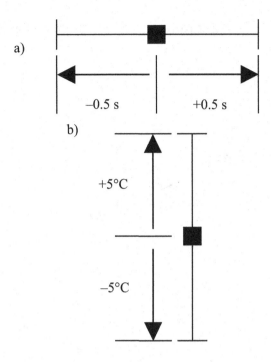

Figure 3.6 (a) Horizontal error bars representing the uncertainty in time. (b) Vertical error bars representing the uncertainty in temperature.

Figure 3.7 shows both the vertical and horizontal error bars attached to all the data points. In this example, the size of the uncertainties does not vary from one point to the next. In some experiments, the uncertainties are not constant from one measurement to the next, such that the size of the error bars varies from point to point on the graph. Where error bars are too small to plot clearly, it is advisable to omit them.

Assuming the y quantity varies smoothly with changes in the x quantity, we can draw a line close to the points as shown in Figure 3.7. In many cases, the y quantity varies smoothly with x quantity (and often it is the 'smooth variation' that we seek to quantify). Consequently, joining the data points one to the next with a straight line is usually discouraged.

In situations where points lie along a curve, it can be difficult to draw a line by hand that passes close to all the points. A strip of stiffened rubber (sometimes referred to as a flexi-curve) is useful when drawing curves on graphs.

It is difficult to extract quantitative information about the relationship between x and y from a curve passing close to the points, the line acting more as a 'guide to the eye'. If quantitative information *is* sought, it is better (if possible) to 'linearise' the data so that a straight line is produced. Straight-line graphs are considered in Section 3.3 and linearisation in Section 3.3.7.

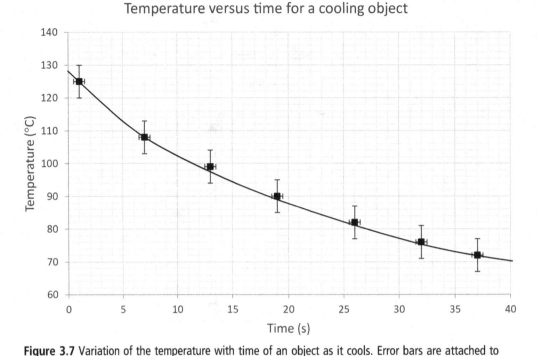

Figure 3.7 Variation of the temperature with time of an object as it cools. Error bars are attached to each point.

3.2.6   When to Plot a Graph

A graph may be plotted either as an experiment proceeds or after the measurements have been completed. Plotting graphs during an experiment can be an efficient use of time and allows for the prompt recognition of interesting or unexpected features in the data.

However, plotting a graph by hand as an experiment proceeds has its disadvantages:

(i) Knowledge is required of the likely range of x and y values; otherwise scales cannot be drawn in advance. Such knowledge is less likely to be available where an experiment is being performed for the first time.

(ii) The experiment may require your careful and sustained attention. The break in concentration needed to plot a data point may cause an important effect to go unnoticed or a critical adjustment to be delayed.

In many situations, it is advisable to attend to ensuring that measurements are made as carefully as possible, leaving graph plotting until the end of the data gathering stage of the experiment.

If data are captured directly with the aid of a computer, graph plotting may be done by the computer as the experiment proceeds. Useful facilities such as the autoscaling of axes are often included in computer-based graph plotting programs, allowing the experimenter to concentrate on other aspects of the experiment.

Once the data points are plotted, they may appear to lie along a curve or a straight line, or show no trend at all. We will look in detail at what is perhaps the most important type of graph: the linear (also known as the straight-line) *x–y* graph.

3.3   Linear *x–y* Graphs

Figure 3.8 shows data obtained from an experiment in which the magnetic field at a point close to a coil of wire (referred to as coil #1) was measured as the electrical current through the wire was increased.

The graph indicates that, as the current increases through the coil, the magnetic field at a point near the coil increases in direct proportion. This is an example of a linear *x–y* graph. The points on the graph do not lie exactly along a straight line. Plausible explanations for the scatter in the points include the possibility of small changes in the current delivered to the coil by the power supply, perhaps due to the electrical resistance of the coil changing as the experiment proceeds. Another explanation could be that currents flowing in nearby equipment are contributing to the magnetic field being measured. Nevertheless, inspection of the graph in Figure 3.8 does appear to support the proposition that there is a linear relationship between magnetic field and current.

A useful tip when trying to decide quickly whether plotted data follow a linear path is to hold the graph paper up to your eyes and continue to tilt the paper until you are looking along the line of the data points. It should be possible to discern whether the data follow a straight line or a line with slight but consistent curvature.

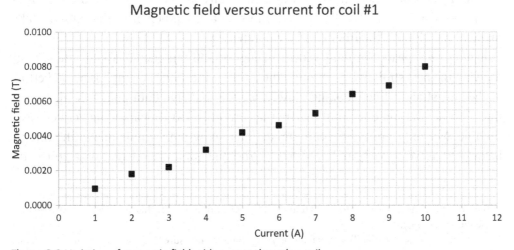

Figure 3.8 Variation of magnetic field with current through a coil.

Linear graphs have an important place in the analysis of experimental data for the following reasons:

(i)   Two important constants can be calculated from a straight line through the points. These are the *slope* of the line[6] and the point at which the line crosses the y axis at $x = 0$, which is referred to as the *intercept*. These will be discussed in Section 3.3.1 and again in Chapter 6.

(ii) Departure from linearity, say at the extremes of the plotted data, can be observed.

(iii) Points that lie far from the line (often referred to as *outliers*) can be identified.

(iv) We can predict values for the y quantity for a chosen value of the x quantity. Likewise, for a particular y value we can use the straight line to find the corresponding value of x.

Let us deal with each of these matters in turn.

3.3.1   The Line of 'Best Fit' through a Set of Data Points

If we are satisfied that a linear relationship exists between the x and y quantities, it is useful to be able to write down an equation that represents that relationship. The first step is to draw a straight line through the points on the graph. As the values gathered through experiment are affected to some extent by sources of variability,[7] it is unlikely that all the points will lie exactly on a straight line even if the underlying relationship between the x and y quantities *is* linear. Therefore, we must use some judgement to estimate the position of the 'best' straight line which passes closest to the points on the graph. This line is often referred to as the 'line of best fit'.

When the line of best fit is drawn by hand, one method for drawing the line through points on an x–y graph is to use a transparent plastic ruler and position it roughly along the line of the points. The next step is to move the ruler until the data points on the graph appear to be scattered evenly above and below the line. It is recommended that the best line be drawn using a sharp pencil so that any mistake made in positioning the ruler can be corrected.

Figure 3.9 repeats the data shown in Figure 3.8, but includes two attempts at drawing the best straight line through the points. Line 1 on the graph is a good effort at the best straight line. Looking closely, we see that five data points lie above the line, four below, and one point (where the current is 4A) appears to be very close to the line. Not only are there approximately as many data points above the line as below, but as we move from left to right across the graph we see that data points are

---

[6] The *slope* of a line is sometimes referred to as the *gradient* of a line (in this book we will consistently use the term *slope*).

[7] Identifying causes of variability in values obtained through measurement is a persistent preoccupation of many scientists and engineers.

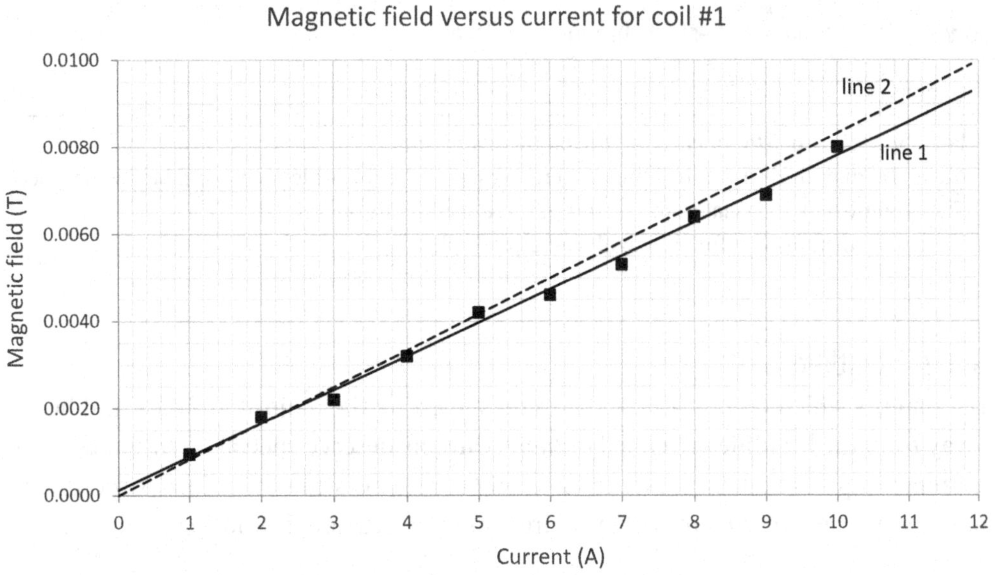

Magnetic field versus current for coil #1

Figure 3.9 A linear *x*–*y* graph with two 'best' lines drawn through the points.

scattered above and below the line. A random distribution of points above and below the line is a distinguishing feature of the best straight line through a set of points.

Line 2 on the graph shown in Figure 3.9 is less satisfactory than line 1 as the best line through the points. The line fits the first few points quite well, but at large values of current it lies consistently *above* the data points. It also appears that line 2 has been deliberately drawn through the origin. All data points should be given equal weight, unless we have some good reason to do otherwise.[8] Consequently, even if the origin *is* a data point, no special effort should be made to force the line through it. If it is believed that the magnetic field should be zero when the current is zero, then we need to investigate why the best line does not go through the origin. It is possible that the instrument measuring the magnetic field has a zero offset which was not taken into consideration,[9] or perhaps there was another source of magnetic field, such as the Earth's magnetic field, that was not accounted for. Whatever the situation, it is unwise to force a line through the origin without a strong reason to support your action. Table 3.4 summarises the steps to take when drawing a best straight line 'by eye'.

Another method for finding the best straight line through a set of *x*–*y* data is considered in Chapter 6.

---

[8] We will deal with situations where experimental data should be 'weighted' in Chapter 6.

[9] Offsets are discussed in Section 4.4.1.

Table 3.4 How to draw the line of best fit through x–y data

Step	Action
(i)	Position a clear plastic ruler along the plotted data points
(ii)	Move the ruler until the points are scattered as equally as possible above and below the line
(iii)	The origin is not a special point so do not force the line through it
(iv)	Using the ruler, draw a fine line with a sharp pencil

3.3.2   Outliers

Data points which do not follow the trend shown by the majority of data and lie far away from the best line are called *outliers*. Outliers deserve special attention because a proper explanation of their presence might require a modification of the theory underlying the experiment. Other, more common, reasons for outliers are:

- incorrect recording of measurements during an experiment
- a mistake made when the data were plotted
- a local disturbance, such as the switching on and off of electrical equipment, which influenced measurements made with an instrument.

Unless the cause of an outlier can be found immediately, the best advice is to repeat the measurement to determine whether the outlying point is reproducible. To throw away a data point because it spoils the look of the graph could be to discard the most important measurement you have made.

3.3.3   Interpolation and Extrapolation

Once the best line has been drawn through a set of x–y data, it is an easy matter to find a y value at any given x value, or vice versa. If we find a value of y for an x value that lies within the range of data points we have plotted, this is termed *interpolation*. For example, with reference to the graph in Figure 3.9, we see that when the current is 7.6 A, the value of magnetic field that corresponds to that current found using the best line (line 1) is 0.0060 T.

When x values lie outside the measurement range, the corresponding y value is found by *extrapolation*. Referring again to Figure 3.9, when the current is 10.8 A, the corresponding magnetic field value is 0.0085 T.

Caution should be exercised when interpolating, and even more so when extrapolating. It may be that the linear relationship breaks down at some value of x beyond the chosen measurement range, so that extrapolation is not valid. Though less likely, some non-linearity could exist *between* consecutive data points, especially if those points are widely separated.

If non-linearity is suspected, then it is sensible to repeat the experiment so that it covers both a greater number and a wider range of x values.

Exercise B

(1) Using line 1 in Figure 3.9, find the value of magnetic field when the current is:
  (i) 5.5 A
  (ii) 11.5 A
(2) A hydrometer is an instrument which indicates the density of a liquid relative to that of pure water.[10] Table 3.5 shows measurements made with a hydrometer for water samples that contain various concentrations of salt.
  (i) Plot a graph of relative density versus salt concentration.
  (ii) Draw the line of best fit through the data.
  (iii) The relative density of a sample of water taken from an estuary is measured using the hydrometer and is found to have a value of 1.030. Use the line on your graph to estimate the salt concentration in the water.

Table 3.5 Variation of relative density with salt concentration

Salt concentration (mg/cm^3)	Relative density (no units)
0	1.005
50	1.034
100	1.066
150	1.095
200	1.122
250	1.150

3.3.4   The Slope and Intercept of the Best Straight Line

The *y* coordinate of any point on a straight line is related to the corresponding *x* coordinate by the equation[11]

$$y = mx + c, \tag{3.1}$$

where *m* is the *slope* of the line and *c* the *intercept*. An accurate determination of *m* and *c* is important as they are parameters that can be compared between various experimenters who are studying the relationship between the same physical

---

[10] The scale on a hydrometer gives the *relative density* (r.d.) of a liquid, defined as r.d. = (density of liquid)/(density of pure water). Being the ratio of two quantities that have the same unit, the relative density itself is a number without units.

[11] It is quite common to see this equation written using other symbols for the slope and intercept, such as $y = mx + b$ or $y = ax + b$ or $y = bx + a$. In this book, we will consistently use the form $y = mx + c$.

quantities. Often, the value of an important physical constant can be calculated using the value of m and/or c. Let us consider a specific example.

The amount that a material expands or contracts when its temperature changes is of importance in the design of structures using that material. Failure to allow for thermal expansion could result in stresses in a structure, causing it to crack or buckle. In this context, it is important to know the coefficient of linear expansion, α, of the material as it permits the length change of a material to be predicted given its original length and the temperature change. If we were to use aluminium as a construction material, we could find α for aluminium by using the graph shown in Figure 3.2. It turns out that (once the best line has been drawn through the points) α is equal to the slope of the line divided by the intercept.

To understand how we can calculate m and c for a straight line, consider Figure 3.10. Two points have been chosen on the line and their coordinates written as (x_1, y_1) and (x_2, y_2).

Using equation 3.1 we have

$$y_1 = mx_1 + c \tag{3.2}$$

$$y_2 = mx_2 + c. \tag{3.3}$$

Solving equations 3.2 and 3.3 to find m gives

$$m = \frac{y_2 - y_1}{x_2 - x_1}. \tag{3.4}$$

$y_2 - y_1$ is sometimes referred to as the *rise* and $x_2 - x_1$ as the *run*, so another way of writing the slope, m, is

$$m = \frac{\text{rise}}{\text{run}}. \tag{3.5}$$

The two points indicated by crosses in Figure 3.11 have coordinates $x_1 = 1$, $y_1 = 4$, $x_2 = 11$ and $y_2 = 25$. Applying equation 3.4, we find that m is

$$m = \frac{25 - 4}{11 - 1} = 2.1.$$

When finding the slope of the line:

(i)   only choose points that lie directly *on* the best line. This normally means that data points cannot be used for determination of the slope.

(ii) choose points on the line that are well separated, as in Figure 3.11, as this lessens the effect of small inaccuracies arising from measuring the run and rise from the graph.

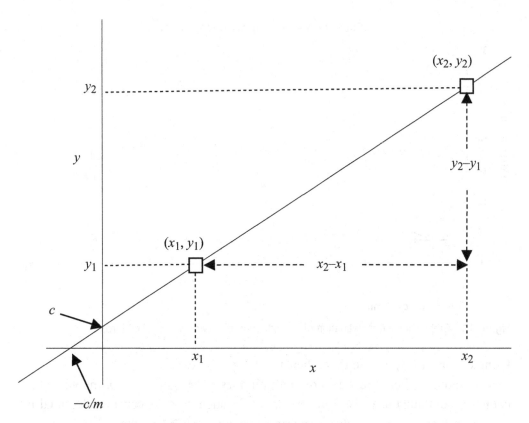

Figure 3.10 Straight line through points on an *x*–*y* graph.

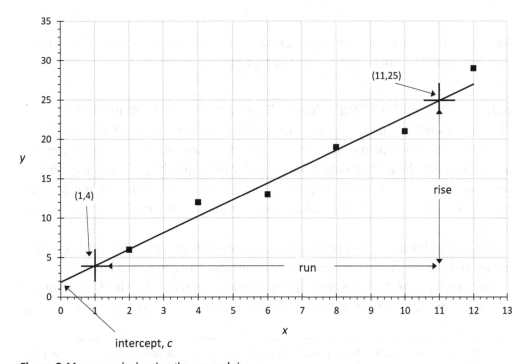

Figure 3.11 *x*–*y* graph showing the run and rise.

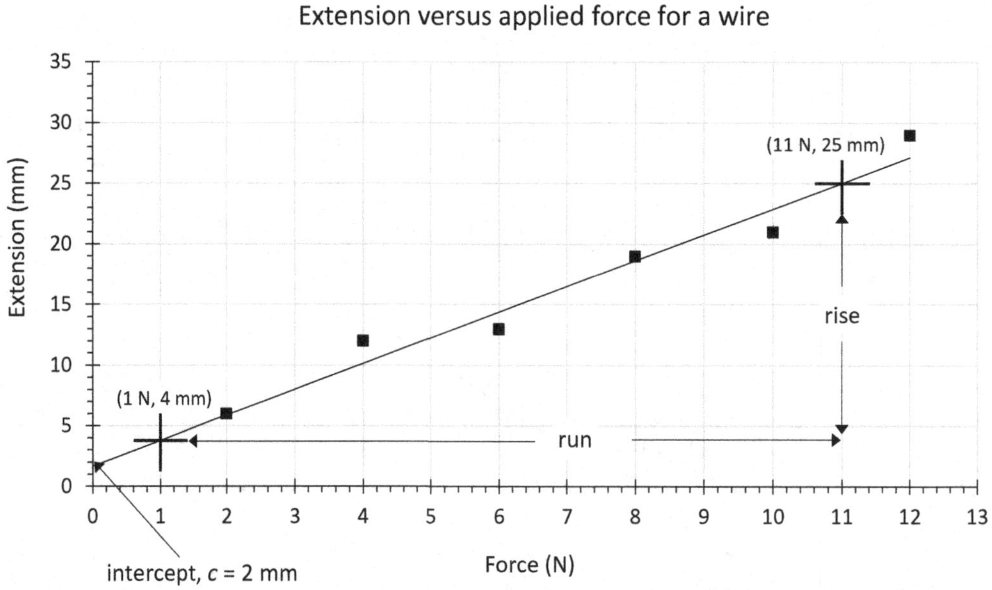

Figure 3.12 Graph showing the variation of the extension of a wire with applied force.

Using equation 3.1, we see that when $x = 0$, then $y = c$.

In situations where the x axis on a graph does not begin at $x = 0$, the value of c cannot be obtained directly from the graph. In such cases, c can be calculated by first finding m as above and then rearranging equation 3.1 so that

$$c = y - mx. \tag{3.6}$$

By choosing any point on the line and substituting the x,y coordinates of that point into equation 3.6, along with m, we can find the value of c.

3.3.5   Slope and Intercept When Dealing with Experimental Data

There are situations, particularly in mathematics, where the points on a graph do not represent experimental data and so the axes have no units of measurement associated with them. By contrast, the majority of graphs in science and engineering show relationships between physical quantities and so *do* have units.[12] The units must be included when calculating the slope and intercept of a line on a graph.

The data in Figure 3.12 came from an experiment in which a force was applied to a wire and the extension of the wire caused by that force was measured.

The intercept has the same units as the quantity on the y axis. In the case of the graph in Figure 3.12, the intercept is 2 mm. The slope also has units, found by using

---

[12] There are occasions when ratios of two quantities, each with the same unit, are plotted on the x or y axis. Such ratios have no units.

equation 3.4. As well as inserting the coordinates into the equation, we include the units, so that

$$m = \frac{(25 - 4)\ \text{mm}}{(11 - 1)\ \text{N}} = \frac{21\ \text{mm}}{10\ \text{N}} = 2.1\ \text{mm/N}.$$

Exercise C

The electrical resistance of a specimen of iron is measured over a temperature range of 40 °C to 90 °C. Table 3.6 shows data gathered in the experiment.

Using the data in Table 3.6, plot a graph of resistance versus temperature, beginning the *x* axis at 40 °C and the *y* axis at 6 Ω. Draw the best line through the points and use equation 3.6 to find the resistance of the specimen at 0 °C. What assumption(s) have you made when finding the resistance at 0 °C?

Table 3.6 Variation of resistance with temperature for a specimen of iron

Temperature (°C)	Resistance (Ω)
40	6.22
50	6.51
60	6.74
70	7.05
80	7.29
90	7.51

3.3.6   Uncertainties in Slope and Intercept

The reason for drawing a straight line as close as possible to the data points is to obtain the best possible estimate of the quantitative relationship between the quantities plotted on each axis. As there is an uncertainty in every value obtained through measurement, there must also be an uncertainty in the slope and intercept of the line drawn through the points. If error bars have been drawn on each point, we can use these to assist in estimating the uncertainty in the slope and intercept. Figure 3.13 shows data gathered in a crystal-growing experiment in which the length of the crystal is plotted against time of growth. Error bars have been attached to each point. As the uncertainty in time is too small to be plotted on this graph, the horizontal error bars are omitted.

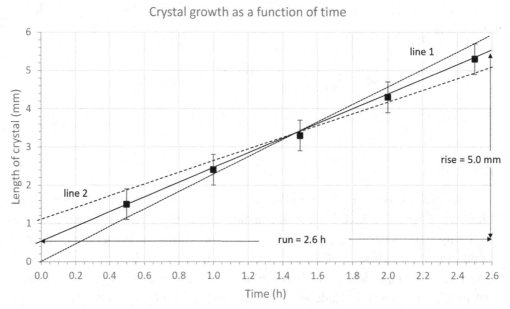

Figure 3.13 Variation of length with time of a growing crystal.

To estimate the uncertainty in the slope and intercept, we draw three lines through the data. The first is the best straight line which passes as close as possible to all the points, as discussed in Section 3.3.1. The slope of this line is[13]

$$\frac{(5.5 - 0.5)\ \text{mm}}{(2.6 - 0.0)\ \text{h}} = \frac{5.0\ \text{mm}}{2.6\ \text{h}} = 1.92\ \text{mm/h}.$$

Lines 1 and 2 are drawn to allow calculation of the maximum and minimum slopes consistent with the error bars. So, for example, line 1 in Figure 3.13 is drawn so that it passes through *all* the error bars, but for the data on the extreme right the line passes close to the top of the error bars and for the points at the extreme left the line passes close to the bottom of the error bars. The slope of line 1 is

$$\frac{(5.8 - 0.0)\ \text{mm}}{(2.6 - 0.0)\ \text{h}} = \frac{5.8\ \text{mm}}{2.6\ \text{h}} = 2.23\ \text{mm/h}.$$

The minimum slope is found using line 2, which passes through the bottom of the error bars for the data points on the extreme right and the top of the error bars for the data on the extreme left. The slope of line 2 is

---

[13] In Figure 3.13 the run and rise are shown for the best straight line. In the interests of clarity, the run and rise for the other two lines have been omitted from the graph.

$$\frac{(5.1-1.1)\ \text{mm}}{(2.6-0.0)\ \text{h}} = \frac{4.0\ \text{mm}}{2.6\ \text{h}} = 1.54\ \text{mm/h}.$$

In this example, the magnitude of the difference between the slope of line 1 and that of the line of best fit is

$$2.23\ \text{mm/h} - 1.92\ \text{mm/h} = 0.31\ \text{mm/h}.$$

The magnitude of the difference between the slope of line 2 and that the line of best fit is

$$1.92\ \text{mm/h} - 1.54\ \text{mm/h} = 0.38\ \text{mm/h}.$$

The uncertainty in the slope can be found by taking the average of these two differences (and rounding), i.e.

$$\frac{(0.31+0.38)\dfrac{\text{mm}}{\text{h}}}{2} = 0.35\ \text{mm/h}.$$

We can now write the slope and its uncertainty as $(1.9 \pm 0.4)\ \text{mm/h}$. We have written the uncertainty to one significant figure and quoted the slope to the same number of decimal places.

To obtain the uncertainty in the intercept, we locate where the three lines cross the y axis. From the graph in Figure 3.13, the intercept of the line of best fit is 0.5 mm. The upper value of the intercept (given by line 2) is 1.1 mm and a lower value (given by line 1) is 0.0 mm. The magnitude of the difference between the intercept of the line 1 and that of the line of best fit is 0.5 mm − 0.0 mm = 0.5 mm. Similarly, the magnitude of the difference between the intercept of line 2 and that of the line of best fit is 1.1 mm – 0.5 mm = 0.6 mm. Taking the average of these two differences gives 0.55 mm, which rounds to 0.6 mm to one significant figure. Now we can give the intercept and its uncertainty as

$$c = (0.5 \pm 0.6)\ \text{mm}.$$

Another method for finding the uncertainties in the slope and intercept is discussed in Section 6.2.3.

3.3.7 Transforming Equations to the Form $y = mx + c$

In many experiments we often have some knowledge of the expected relationship between the quantities being studied. In such a case, and if the relationship can be expressed as an equation, it is sometimes possible to choose what to plot on each axis to produce a straight-line graph. We begin by transforming the equation, where that is possible, into the form $y = mx + c$. Let us consider a specific example.

A body of mass M is attached to one end of a spring, the other end being fixed. The mass is set oscillating vertically. The period of the body is measured as the mass of the body is increased. The graph in Figure 3.14 shows how the period varies with mass.

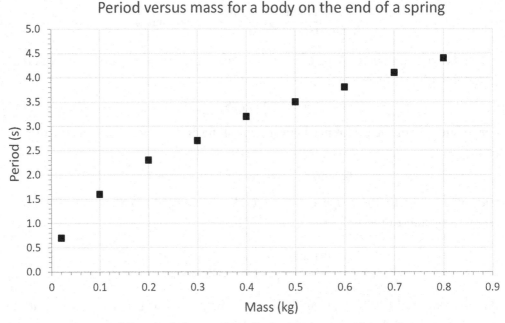

Figure 3.14 Variation of the period of an oscillating body with the mass of the body.

Looking along the points in Figure 3.14, it is clear that the relationship between period and mass is not linear. If, instead of plotting period versus mass, we plot period versus $(mass)^{1/2}$, the outcome is the straight-line graph shown in Figure 3.15.

The graph was linearised by changing the quantity plotted on the x axis. Why did plotting $(mass)^{1/2}$ on the x axis linearise the graph? To understand this, we must look again at the equation of a straight line, $y = mx + c$, and compare this to the equation describing the relationship being studied.

The equation which relates the period of oscillation, T, to the mass, M, on the end of a spring is

$$T = 2\pi\sqrt{\frac{M}{k}},$$

where k is a constant, called the spring constant. To make comparison with $y = mx + c$ easier, the equation can be rewritten with $y = mx + c$ below it, as follows:

$$T = \frac{2\pi}{\sqrt{k}}\sqrt{M}$$

$$y = m \quad x + c$$

(3.7)

We see that if T is plotted on the y axis and $\sqrt{M}$ is plotted on the x axis, the slope of the line is $2\pi/\sqrt{k}$. As no constant appears as a separate term on the right-hand side of equation 3.7, the intercept, c, is equal to zero.

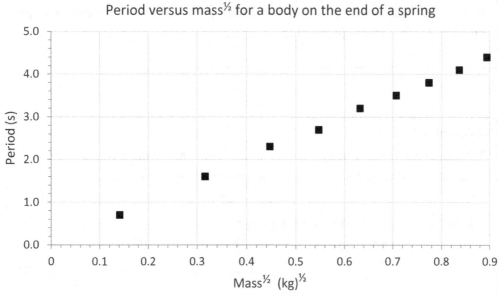

Figure 3.15 Period of a body attached to a spring plotted against (mass)^{½}.

There are a large number of situations where linearisation is possible, but the process of transforming or arranging an equation into the form $y = mx + c$ takes practice, so we will consider three representative examples.

Example 1

In an experiment to study the variation of length with temperature for a metal rod, the equation relating length, l, to temperature, θ, is

$$l_\theta = l_0(1 + \alpha\theta). \tag{3.8}$$

l_θ is the length of the rod at the temperature θ, l_0 is the length at 0 °C and α is the temperature coefficient of linear expansion.

In order to write equation 3.8 in the form $y = mx + c$, the first step is to expand the right-hand side of the equation by multiplying through by l_0, giving

$$l_\theta = l_0 + l_0\alpha\theta.$$

Now we compare this equation with $y = mx + c$. As we have done before, we will write $y = mx + c$ below the equation, rearrange the order of the right-hand side of the equation, then group together corresponding terms in the equations as shown below:

$$l_\theta = l_0\alpha\ \theta + l_0$$

$$y = m\ x + c$$

In this example, plotting l_θ versus θ should give a straight line with a slope equal to $l_0\alpha$ and intercept equal to l_0.

Example 2

The displacement of a body, s, accelerating uniformly with acceleration a is measured at various times, t. The relationship between s and t is

$$s = ut + \tfrac{1}{2}at^2, \tag{3.9}$$

where u is the initial velocity of the body.

In order to linearise equation 3.9, divide throughout by t, then rearrange, to give

$$\frac{s}{t} = \tfrac{1}{2}a\,t + u. \tag{3.10}$$

Proceeding as before, we can compare the terms in equation 3.10 with $y = mx + c$:

$$\frac{s}{t} = \tfrac{1}{2}a \quad t \quad + \quad u$$

$$y \;=\; m \quad x \;+\; c$$

Plotting s/t versus t produces a straight line with slope $\tfrac{1}{2}a$ and intercept u.

This example differs from the previous one in that the independent variable (in this case the time, t) appears on both sides of the equation. In most cases the independent variable appears only on the right-hand side of the equation. However, some situations (such as this one) require this rule to be relaxed.

Example 3

The equation

$$N = N_0 e^{-\lambda t} \tag{3.11}$$

relates the number of undecayed nuclei, N, that remain in a sample of radioactive material after a time t has elapsed. N is the dependent variable, t is the independent variable. N_0 and λ are constants.

To transform an equation which contains a number raised to a power into the form $y = mx + c$, the first step is to take the logarithms of both sides of the equation. In this example, the number raised to the power $-\lambda t$ is the constant, e (which has a value, to seven significant figures, of 2.718282). If we take logarithms to the base e of both sides of equation 3.11 and rearrange, we obtain[14]

$$\ln(N) = -\lambda\, t + \ln(N_0).$$

$$\underset{y}{\downarrow} \quad = \quad \underset{m}{\downarrow} \quad \underset{x}{\downarrow} \quad + \quad \underset{c}{\downarrow}$$

Plotting $\ln(N)$ versus t produces a straight line with slope equal to $-\lambda$ and intercept equal to $\ln(N_0)$.

Exercise D

Table 3.7 contains equations taken from various fields in science and engineering. For each equation the dependent variable, the independent variable and the constants in the equation are indicated.

For each of the equations in Table 3.7:

(i) What would you plot in order to obtain a straight line?

(ii) How are the slope and intercept related to the constants in the equation?

Table 3.7 Equations to be rearranged in the form $y = mx + c$

Equation number	Equation	Dependent variable	Independent variable	Constant (s)
1	$F = \mu N$	F	N	μ
2	$v = u + at$	v	t	u, a
3	$R = AT + BT^2$	R	T	A, B
4	$I = I_0 \exp(-\mu x)$	I	x	I_0, μ
5	$T = 2\pi\sqrt{l/g}$	T	l	g
6	$\dfrac{1}{u} + \dfrac{1}{v} = \dfrac{1}{f}$	v	u	f
7	$H = C(T - T_0)$	H	T	C, T_0
8	$I = AV \exp(-BV^2)$	I	V	A, B

---

[14] To indicate taking logarithms to the base e of the number N, we write $\ln(N)$.

3.4   Logarithmic Graphs

The graphs we have considered so far have been plotted using axes with linear scales on both axes. There are circumstances, however, where no matter how careful we are when choosing such scales, useful information is obscured. As an example of this, consider Table 3.8, which contains data from a study of the electrical characteristics of a silicon diode. More specifically, data are presented of the current through the diode for particular values of voltage across the diode. A graph of the data in Table 3.8, plotted using linear scales is shown in Figure 3.16.

The current data shown in Figure 3.16 span so many orders of magnitude that, in choosing a scale to fit the large values of current on the graph, the values of currents measured at voltages less than 0.6 V cannot be read from the graph. Data spanning many orders of magnitude are better accommodated by replacing linear scales on one or both axes by *logarithmic* scales. Figure 3.17 shows the data from Table 3.8, but here equal distances on the vertical scale correspond to changes by *powers of 10*.

Figure 3.17 shows the current versus voltage data plotted on a log–linear scale.[15] By this we mean that one axis has a logarithmic scale (in this example, the y axis) and the other scale is linear.

Although in this example plotting the quantities on a log–linear scale has produced points which lie in a straight line (indicating that there is a logarithmic relationship between current and voltage), in general a straight line cannot be expected.

Another way to approach the plotting of data that covers many orders of magnitude is to take the logarithms of those data and plot them on conventional

Table 3.8 Current variation with applied voltage for a silicon diode	
Voltage (V)	Current (A)
0.35	9.0×10^{-7}
0.40	3.0×10^{-6}
0.45	5.0×10^{-5}
0.50	2.0×10^{-4}
0.55	1.7×10^{-3}
0.60	1.5×10^{-2}
0.65	7.5×10^{-2}
0.70	0.55
0.75	3.5

---

[15] This is sometimes referred to as a semi-log scale.

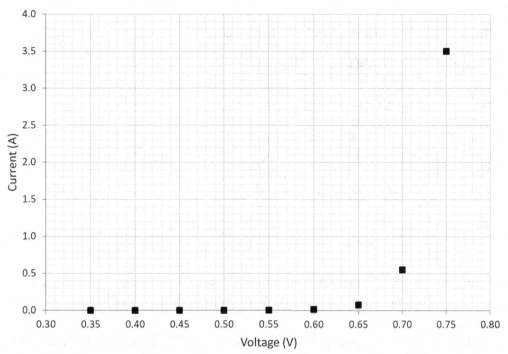

Figure 3.16 Current variation with voltage for a silicon diode. The *x* and *y* scales are both linear.

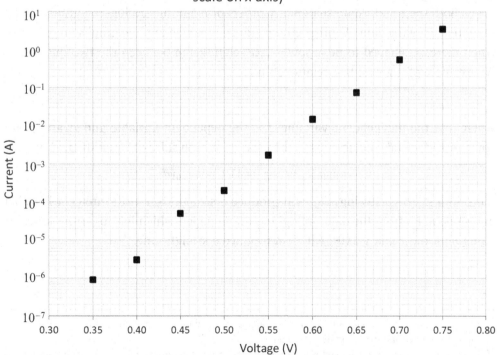

Figure 3.17 Current variation with voltage for a silicon diode. The *y* scale is logarithmic and the *x* scale is linear.

Table 3.9 Current–voltage data for a silicon diode		
Voltage (V)	Current (A)	$\text{Log}_{10}(\text{current(A)})$
0.35	9.0×10^{-7}	-6.046
0.40	3.0×10^{-6}	-5.523
0.45	5.0×10^{-5}	-4.301
0.50	2.0×10^{-4}	-3.699
0.55	1.7×10^{-3}	-2.770
0.60	1.5×10^{-2}	-1.824
0.65	7.5×10^{-2}	-1.125
0.70	0.55	-0.260
0.75	3.5	0.544

linear graph paper.[16] It does not matter to which base we take the logarithms, but by far the most common method is to take logarithms to the base e or logarithms to the base 10.

Table 3.9 shows the current and voltage data for the diode given in Table 3.8 with the addition of a column giving the logarithms of the current values to the base 10. A plot of $\log_{10}(\text{current(A)})$ versus voltage is shown in Figure 3.18.

3.4.1 Log–Log Graphs

When both the x and y data span many orders of magnitude, a log–log graph can be used to display the data. Another circumstance in which log–log graphs are useful is when there is believed to be a power-law relationship between the two quantities. An example of a power law relationship is

$$I = Ad^n, \tag{3.12}$$

where d is the independent variable, I is the dependent variable and A and n are constants. If we take logarithms to the base 10 of both sides of equation 3.12, we get

$$\log_{10}(I) = \log_{10}(A) + n\log_{10}(d).$$

Plotting $\log_{10}(I)$ versus $\log_{10}(d)$, will produce a straight line with intercept $\log_{10}(A)$ and slope n. Plotting the data on log–log graph paper, which has logarithmic scales along *both* the x and y axes, will also produce a straight line.[17]

---

[16] The convenience and availability of linear graph paper often makes this method of plotting data more attractive than using log–linear paper.

[17] Log–log graph paper is available for plotting data covering many orders of magnitude but, just as for log–linear graphs, it is often more convenient to use linear paper after taking the logs of the quantities.

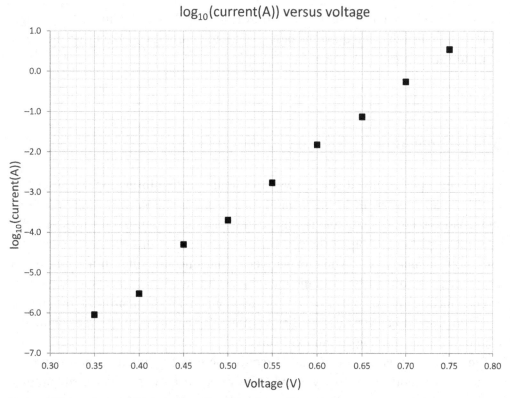

Figure 3.18 $\log_{10}$(current(A)) variation with voltage for a silicon diode.

Exercise E

Table 3.10 contains data on the amount of energy, H, emitted from a hot body each second over a range of temperature, T. It is believed that there is a power law

Table 3.10 Energy emitted from a body at temperatures from 1500 K to 2500 K	
T (K)	H (W)
1 500	150
1 600	190
1 700	230
1 800	300
1 900	360
2 000	440
2 100	560
2 200	680
2 300	800
2 400	930
2 500	1 100

relationship between H and T of the form $H = AT^n$. Plot a log–log graph using the data in the table and draw the best straight line through the data. Use the best line to find values for A and n.

3.5   Comment

Graphs are a powerful and concise way to communicate information. Representing data from an experiment in the form of an x–y graph allows us to study relationships, assess scatter in data and rapidly identify special or unusual features. A well-laid-out graph containing all the components discussed in this chapter can act as a 'one stop' summary of a whole experiment. Someone studying an account of an experiment will often first examine the graph(s) included in the account to gain an overall picture of the outcome of the experiment. The importance of graphs, therefore, cannot be overstated as they so often play a central role in the communication of the key findings of an experiment.

Programs such as Excel or Graph[18] reduce the effort required to plot graphs as compared to plotting them by hand. Nevertheless, even sophisticated plotting programs can present data in a manner that is not easy to absorb. For example, when using a plotting program's default settings, it is possible to create x–y graphs that have poorly chosen scales on the axes and omit important information such as units of measurement, or that include the origin, causing data to be confined to a small region on the graph. Fortunately, most plotting programs contain versatile plotting options, and allow us to choose:

- the way each axis is labelled
- the shape and size of markers used to represent data points
- the scales on each axis
- whether or not the origin appears on the graph.

We will consider graph plotting using Excel in Chapter 8.

---

[18] Graph is a free application for drawing x–y graphs. It is available from http://www.padowan.dk/

Problems

3.1. Table 3.11 shows the pressure, P, of a fixed volume of gas measured as the temperature, T, of the gas is increased.

Plot a graph of pressure versus temperature and draw the best straight line through the points. Find:

  (i) the slope of the line

  (ii) the intercept of the line (i.e. the value of pressure corresponding to $T = 0\,\mathrm{K}$)

  (iii) the pressure at 330 K and 400 K

  (iv) the temperature corresponding to a pressure of $1.00 \times 10^5\,\mathrm{Pa}$.

3.2. For each of the equations in Table 3.12:

  (i) what would you plot in order to obtain a straight line?

  (ii) how are the slope and intercept related to the constants in the equation?

Table 3.11 Pressure of a fixed volume of gas as a function of temperature

$P\,(\times\,10^5\,\mathrm{Pa})$	$T\,(\mathrm{K})$
0.92	280
0.98	295
1.02	305
1.05	320
1.15	350
1.20	370
1.30	390

Table 3.12 Equations to rearrange into the form $y = mx + c$

Equation number	Equation	Dependent variable	Independent variable	Constant(s)
1	$T_w = T_c - kR^2$	T_w	R	T_c, k
2	$D = \tfrac{1}{2}CA\rho v^2$	D	v	C, A, ρ
3	$\dfrac{1}{f} = \dfrac{n-1}{r}$	f	r	n
4	$hf = \phi + E$	E	f	h, ϕ
5	$t = \dfrac{A}{D^2}(1 + BD)$	T	D	A, B
6	$E = E_0\left(1 + \dfrac{v}{c}\cos(\theta)\right)$	E	θ	E_0, v, c
7	$\Delta T = k\log_{10}A + D$	ΔT	A	k, D

3.3. A rod suspended from two strings forms a bifilar pendulum, as shown in Figure 3.19. The strings are separated by a distance d, and are equidistant from the ends of the rods.

The rod is made to oscillate in a horizontal plane about the axis O. The period of the pendulum, T, is measured as the separation, d, between the strings is varied. Table 3.13 contains the period versus separation data.

Assume the relationship between T and d can be written

$$T = kd^a. \tag{3.13}$$

(i) Transform equation 3.13 into the form $y = mx + c$.

(ii) Based on the transformed equation, plot a suitable graph and draw the best straight line through the data points. Use your best line to determine k and a.

(iii) Use your best line to estimate the period when $d = 45$ cm.

3.4. In an optics experiment, an image of an illuminated object is formed using a lens. The distance from the object to the lens is u and the distance from the lens to the image is v. The experiment consists of varying u and measuring the corresponding v. Table 3.14 contains the data gathered in the experiment.

Table 3.13 Variation of the period of bifilar pendulum, T, with string separation, d

d (cm)	T (s)
80	0.88
70	1.01
60	1.17
50	1.41
40	1.75
30	2.32
20	3.52
10	7.01

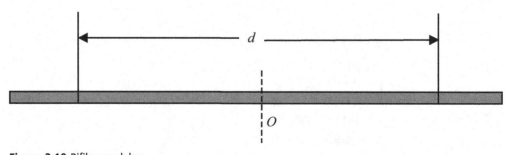

Figure 3.19 Bifilar pendulum.

Table 3.14 Object and image distance data for a lens	
u (cm)	v (cm)
27.0	59.0
31.0	46.2
36.1	38.5
49.8	29.5
59.5	26.9
71.2	25.2
84.8	23.1

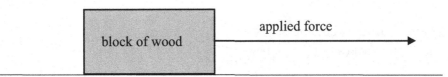

Figure 3.20 Force applied to a block of wood on a metal surface.

Assuming the relationship between u and v is $1/u + 1/v = 1/f$, where f is a constant called the focal length of the lens:

(i) rearrange the equation into the form $y = mx + c$

(ii) based on the rearranged equation, plot a suitable graph and draw the best straight line through the data points

(iii) use your best line to determine f.

3.5. In an experiment to study the frictional forces between two bodies, a force was applied to a block of wood situated on a flat, smooth metal surface, as shown in Figure 3.20. Table 3.15 contains data showing how the minimum force required to move the block varies with the mass of the block.

(i) Plot a graph of minimum force against mass. Attach error bars to each point.

(ii) Draw the best straight line through the points and calculate the slope and intercept of the line.

(iii) Write the equation of the line in the form $y = mx + c$.

(iv) Using the equation of the line, estimate the minimum force required to move the block, when the mass of the block is:
(a) 0.70 kg
(b) 1.30 kg

(v) With the aid of the error bars, estimate the uncertainty in the slope and intercept.

Table 3.15 Minimum applied force required to move the block as the mass of the block increases

Mass (kg) (± 0.01 kg)	Minimum force (N) (± 0.2 N)
0.52	3.1
0.58	3.6
0.64	3.9
0.75	4.4
0.88	5.2
1.01	6.2
1.21	7.4

Table 3.16 Light absorption as a function of mercury concentration

Mercury concentration (parts per billion)	Absorption (arbitrary units)
0.5	10
1.0	20
2.0	38
3.0	56
4.0	75

3.6. The electrical current, I, flowing around a circuit was measured as the resistance in the circuit, R, was varied. The relationship between I and R is given by

$$I = \frac{E}{R + r},$$ (3.14)

where E is the e.m.f. of a battery in the circuit and r is its internal resistance (E and r are constants).

  (i) Transform equation 3.14 into the form $y = mx + c$ and state what you would plot on the x and y axes in order to obtain a straight line.

  (ii) State how E and r are related to the slope and/or the intercept of the graph.

3.7. An experiment was performed to determine the amount of mercury in river water using atomic absorption spectrophotometry. Calibration data of light absorption as a function of mercury concentration (in parts per billion, or ppb) are shown in Table 3.16.

  (i) Plot a graph of absorption versus concentration.

  (ii) Draw the best straight line through the points.

(iii) A sample of river water gives an absorption of 25 ± 3 arbitrary units. Use the best line on your graph to estimate the concentration of mercury in the river water and the uncertainty in the concentration.

3.8. A ceramic conductor was observed to exhibit unusual electrical characteristics at low temperature. Table 3.17 contains data on the electrical resistance of a sample of the ceramic as the current through it is increased.

Assume that the relationship between R and I can be written

$$R = kI^n,$$

where k and n are constants.

(i) Plot a suitably linearised graph of the data given in Table 3.17.
(ii) Draw the best straight line through the points.
(iii) Calculate the slope and intercept of the line.
(iv) Use the slope and intercept to find values for k and n.

3.9. When sodium thiosulfate ($Na_2S_2O_3$) is mixed with hydrochloric acid, a reaction occurs during which a sulfur precipitate is formed. An experiment was performed to study the rate of reaction as a function of concentration of $Na_2S_2O_3$. Table 3.18 shows the time for the reaction to reach a specific point as a function of the concentration of $Na_2S_2O_3$.

It is believed that the reaction rate for these chemicals should depend linearly on the concentration of $Na_2S_2O_3$ present.

(i) Taking the reaction rate to be equal to 1/(reaction time), draw up a table of reaction rate as a function of concentration.
(ii) Plot a graph of the reaction rate versus concentration.
(iii) Draw the best straight line through the points and calculate the slope of the line.

Table 3.17 Resistance of a ceramic conductor as a function of current

I (A)	R (Ω)
1.0×10^{-3}	6.0×10^{-4}
2.0×10^{-3}	2.2×10^{-3}
4.0×10^{-3}	6.3×10^{-3}
8.0×10^{-3}	2.0×10^{-2}
1.6×10^{-2}	4.2×10^{-2}
3.2×10^{-2}	1.2×10^{-1}
6.4×10^{-2}	3.4×10^{-1}
1.3×10^{-1}	1.1
2.6×10^{-1}	3.2
5.2×10^{-1}	9.5

Table 3.18 Time for reaction of $Na_2S_2O_3$ and hydrochloric acid as a function of concentration of $Na_2S_2O_3$

Concentration (mol)	Reaction time (s)
0.20	25
0.16	30
0.14	32
0.12	40
0.08	56
0.02	116

4 Dealing with Uncertainties

4.1 Overview: What Are Uncertainties?

An experiment may require that measurements be made with simple equipment such as a stopwatch or a metre ruler, or it may involve sophisticated instruments such as those carried by NASA's Mars Curiosity Rover, designed to analyse rock, soil and air samples. Through measurement we are trying to determine the value of a quantity such as the time for an object to fall a fixed distance to the ground, the distance between a lens and an image formed by the lens or the chemical composition of minerals found on Mars.

If we were to make repeat measurements of a quantity, we would likely find that the values obtained would vary one to the next. This leads us to the idea that there is an amount of *uncertainty* in values obtained through measurement. In this chapter we will look at ways of recognising and dealing with uncertainties arising from measurement.

4.1.1 Example of Variability in Values Obtained through Measurement

Consider an experiment in which a small object falls through a fixed distance and the time for it to fall is measured using a stopwatch. Table 4.1 contains 10 values recorded for the time of fall.

We might have hoped that, on each occasion when we measured the time of fall of the object, we would obtain the same value. This is not true in this experiment, and in general it is untrue of *any* experiment.[1] We must acknowledge that variability in values obtained through measurement is an inherent feature of all experimental work. What we need to be able to do is recognise, examine and quantify the variation; otherwise the reliability of our experiment may be questioned, and any conclusions drawn from the experiment may be of limited value. If it is possible to identify the main cause(s) of the variation in the experimental data, then we may be able to redesign the experiment to reduce the variability.

---

[1] There is an exception to this. When items are counted (for example the number of pages in this book) it is possible to do this with complete accuracy.

Table 4.1 Times for object to fall a fixed distance									
Time (s)	0.64	0.61	0.63	0.53	0.59	0.65	0.60	0.61	0.64	0.71

4.1.2 True Value

Perhaps we could determine the exact time for a ball to fall a fixed distance described in the example in Section 4.1.1, if we could:

- devise a means for releasing the ball in the exact same manner every time
- perfectly synchronise the start of a clock with that release
- ensure that the distance that the ball falls before it hits the ground never varies
- establish precisely the moment when the ball touches the ground
- eliminate possible interfering effects such as the local movements of air around the ball
- use a clock that measures time intervals exactly, which would require, amongst other things, that it had infinitely fine resolution.

If these conditions were satisfied, we might be able to claim to have found the 'true value' of the time of fall of the ball. The reality is that these conditions cannot be met. For example, the best clocks, though very good, have finite resolution. While an excellent mechanism can be engineered to release the ball, there will always be tiny variations in the way a ball is released. In short, though we can speak of a true value for the time of fall of the ball, we can never know it. This is also the case for the value of any other quantity we care to establish through measurement.

4.1.3 Error

Scientists and engineers speak of the 'error' in a value and seek ways to reduce the error. In this case the word error does not refer to a mistake or blunder that might have been made, for example by incorrectly recording a value indicated by an instrument, or wrongly converting a velocity in km/h to m/s. In the context of an experiment, an error is the difference between the value of that quantity obtained through measurement and the true value of a quantity. We can write this as an equation:

$$\Delta x = x - \mu, \tag{4.1}$$

where μ is the true value of a quantity, x is the value of that quantity obtained through measurement and Δx is the error. As μ cannot be known, it follows that the error, Δx, is also unknowable.

What we *can* do is establish a best estimate of a quantity, then use our knowledge, for example of the limitations of the instrument or the variability we observe when we make repeat measurements, to determine the *uncertainty* in the best estimate. The uncertainty will allow us to define a range within which we can confidently expect the true value of the quantity to lie.

In summary: the error in values obtained through measurement cannot be known, but we are able to determine the uncertainty in the best estimate of the true value.[2]

Exercise A

(1) Fill an electric kettle with an amount of cold water (say 500 cm^3). Switch the kettle on and measure the time it takes for the water to reach boiling point. Empty the kettle completely and repeat the experiment twice. What values did you obtain for the boiling times? What factors do you think are responsible for the variation in times?

(2) If your wristwatch or phone has a stopwatch option, take a small coin and let it fall from a fixed height (say 3 m). Measure the time it takes to fall to the ground. Repeat the experiment 9 times. What are the minimum and maximum times? What do you think are the most important factors affecting the timings?

4.2 Uncertainty in a Single Measurement

In any experiment it is easy to make a mistake while carrying out a measurement. If you have made only one measurement of a quantity, you need to be confident that it is reliable; otherwise a spurious value may affect the success of your whole experiment. It is much better to make the measurement at least once more to be satisfied that the value obtained is repeatable.[3] There are circumstances in which it is difficult to make repeated measurements of a quantity, for example when the quantity is continually changing with time. In these situations, you may have to accept that a single measurement is the best that you can do. In the following sections we discuss what factors affect the uncertainty you would quote in a value obtained through a single measurement.

---

[2] In some books you will find the words 'error' and 'uncertainty' used interchangeably. In this book we will consistently use the descriptions given in Section 4.1.3.

[3] Suppose we make repeat measurements of a quantity under conditions that remain unchanged, i.e. the same observer, location, equipment, measurement method and environmental conditions. If the scatter in the values obtained through measurement is small, then we say that the measurement is *repeatable*.

4.2.1   Resolution Uncertainty

Every instrument has limited resolution. Take, for example, the measurement of length using a metre ruler. The finest markings on the scale of such a ruler are usually separated by 1 mm. You might be able, if you were careful, to measure a length to within 0.5 mm, but it is unlikely that you could do much better than this. If you required a 'better' measurement of length, you could use vernier calipers or a micrometer which (typically) are capable of resolutions of 0.1 mm and 0.01 mm, respectively. There are other length measuring instruments, based on lasers, that offer even better resolution. Nonetheless, all measurements are limited by the instrument you are using. If the quantity you are measuring is stable or at least varies slowly with time, it is reasonable to quote the uncertainty as one-half of the smallest division on the scale. For example, you could quote a length measured with a metre ruler as (361.0 ± 0.5) mm. The $\pm$ sign is used as shorthand to indicate that the length lies somewhere in the range 360.5 mm to 361.5 mm. In the case of a thermometer whose smallest division is 2 °C, a temperature could be (61 ± 1) °C. In general, the resolution limit of an instrument represents the smallest uncertainty that can be quoted in a single measurement of a quantity.

4.2.2   Reading Uncertainty

While making a measurement it is possible that the quantity under investigation varies by considerably more than one-half of the smallest division on the instrument. Imagine heating a beaker containing water. We may be using a thermometer that has a resolution of 1 °C or even better. However, we notice that, as the water is stirred (in an effort to ensure a uniform temperature throughout), the thermometer indicates a wide temperature variation. At one instant the thermometer indicates 36 °C, at the next 32 °C, then quickly changes to 35 °C. Quoting an uncertainty of ± 1 °C in these circumstances would be to underestimate the uncertainty. In this situation we must use some judgement to decide a reasonable value for the uncertainty. In a one-off measurement of this kind, there are no hard and fast rules about quoting uncertainties, and we have to rely on our common sense. If we estimate the reading uncertainty to be less than ± 5 °C, but greater than ± 1 °C, then we should choose a compromise between these two values.

4.2.3   Calibration Uncertainty

Instruments used in industrial, research or analytical laboratories are routinely calibrated against highly accurate standards (a standard has a value with very small uncertainty). For example, accurate weights are used to calibrate laboratory balances and scales. Expensive measuring equipment is often supplied with a calibration certificate indicating how closely the instrument conforms to the standard. If scientists around the world are trying to compare their measurements, they need to be sure that their instruments agree on what is a metre, a volt, a second and so on.

The calibration certificate provided by a manufacturer or a bureau of standards is unlikely to guarantee that the calibration is valid for more than one year. For the calibration to remain valid, the instrument must be checked regularly.

It is unlikely that you will have the time or facilities to make a thorough check of the calibration of instruments you are using. However, a quick comparison of, say, voltages indicated by two voltmeters measuring the same voltage, or two thermometers which should be indicating the same temperature, may save you from having to repeat a whole experiment because of a poorly calibrated, or perhaps faulty, measuring instrument. A poorly calibrated instrument leads to *systematic* error in values and influences all measurements made with that instrument. We will have more to say on systematic error in Section 4.4.

It is important to be aware of resolution, reading and calibration uncertainties when attempting to quote the uncertainty in a single measurement. Such uncertainties exist whenever a measurement is made in an experiment. However, to be able to better quantify the variability in values obtained through measurement, more than one measurement should be made of each quantity. Where this is possible we can use some results from statistical analysis to allow us to quantify uncertainties. Let us turn to arguably the most important quantity that can be calculated from a series of repeated measurements: the *mean.*

4.3   The Best Estimate of a Quantity Obtained through Repeat Measurements: The Mean

Look again at the values given in Table 4.1 for the time of fall of an object. We could expect the time that it really took for the object to fall to lie somewhere between the two extreme measured values, namely between 0.53 s and 0.71 s. If a single value for the time of fall is required, we can do no better than to calculate the average of the 10 measurements that were made.

The average of a group of numbers is termed the *mean* of the numbers. The symbol used for the mean is $\bar{x}$, and it is calculated using

$$\bar{x} = \frac{\sum_{i=1}^{i=n} x_i}{n}, \tag{4.2}$$

where n is the number of measurements made. Σ is a summation sign which instructs us to add the x's together, so that[4]

$$\sum_{i=1}^{i=n} x_i = x_1 + x_2 + x_3 + \cdots + x_n.$$

---

[4] x_1 refers to the first value, x_2 the second value and so on.

Using the experimental data given in the Table 4.1, the mean time is given by

$$\bar{x} = \frac{(0.64 + 0.61 + 0.63 + 0.53 + 0.59 + 0.65 + 0.60 + 0.61 + 0.64 + 0.71) \text{ s}}{10}$$

$$= \frac{6.21 \text{ s}}{10} = 0.621 \text{ s}.$$

We could quote the mean to one, two or three significant figures, i.e. 0.6 s, 0.62 s or 0.621 s. Which do we choose? This question is difficult to answer unless we have an estimate for the uncertainty in the best estimate. We will not the use the methods we introduced in Section 4.2 to estimate the uncertainty in a single measurement. Instead we will use the variability in the gathered data as a guide to estimating the uncertainty.

Exercise B

(1) In a fluid flow experiment, the amount of water flowing through a pipe in a fixed time is measured. Given in Table 4.2 are measured volumes of water passing through the pipe, collected over eight successive 2-minute periods. Calculate the mean volume collected in cm^3, and the mean rate of flow in m^3/s.

(2) The gas pressure in the specimen chamber of an electron microscope was recorded every 10 minutes over a period of 2 hours. The data are presented in Table 4.3. Calculate the mean pressure over the 2-hour period.

4.3.1   Uncertainty in the Best Estimate

We now consider a method by which the uncertainty in the best estimate can be calculated. In Chapter 5 we will deal in more detail with estimating uncertainties in data gathered, and discuss another method by which the uncertainty can be calculated.

Table 4.2 Volume of water flowing through a pipe

Volume (cm^3)	48	45	45	50	47	46	43	44

Table 4.3 Gas pressure in the specimen chamber of an electron microscope

Pressure ($\times 10^{-3}$ Pa)	3.2	7.5	8.0	6.5	1.0	0.9
	4.4	5.0	4.0	5.5	2.3	3.4

In situations where, say, between 5 and 10 repeat measurements have been made of a quantity, an estimate of the uncertainty can be made by first calculating the *range* of the data, given by

$$\text{range} = \text{largest value} - \text{smallest value}.$$

The uncertainty in the best estimate[5] may be found by dividing the range by the number, n, of measurements that were made:[6]

$$\text{uncertainty} = \frac{\text{range}}{n}. \tag{4.3}$$

As an example of the application of this method for calculating uncertainty, consider an experiment in which eight measurements were made of the speed of sound in air at 20 °C. The recorded values are shown in Table 4.4. The mean of these values is 341.775 m/s and the range is 345.5 – 338.5 = 7 m/s. Using equation 4.3, we find that the uncertainty is 7/8 = 0.875 m/s.

Exercise C

(1) Using the values in Table 4.1, calculate the uncertainty using equation 4.3.
(2) In an experiment to study the properties of convex lenses, the distance from a lens to an image was measured on six occasions. The values obtained are given in Table 4.5. Calculate the mean value of the image distance and the uncertainty using equations 4.2 and 4.3.

Table 4.4 Experimental values of the speed of sound in air at 20 °C

Speed of sound (m/s)	341.5	342.4	342.2	345.5	341.1	338.5	340.3	342.7

Table 4.5 Values of image distance as measured for a convex lens

Image distance (mm)	425	436	428	417	429	413

[5] For a justification of equation 4.3, see Lyon (1980).
[6] In Section 5.3 we will consider another method of calculating the uncertainty which gives values consistent with those obtained using equation 4.3.

4.3.2   How to Quote Uncertainties

In the example in Section 4.3.1, we might be tempted to say that the speed of sound in air, based upon the measurements we have made, is 341.775 m/s with an uncertainty of 0.875 m/s. We must be careful when we quote uncertainties. Their function is to quantify the interval in which the true value of that quantity probably lies.[7] There is no point, therefore, in quoting the uncertainty to more than one or two significant figures, and usually one significant figure will suffice.[8]

In the present example we would round 0.875 m/s up to 0.9 m/s. Such a rounding indicates that the mean itself should not be quoted, as we have done, to three decimal places. In this situation we round the mean to the same number of decimal places as the uncertainty.

In this example we would round 341.775 m/s to 341.8 m/s. We can say that, based on data from the experiment, the value for the speed of sound (symbol, v) in air is 341.8 m/s with an uncertainty of 0.9 m/s. This is rather a long-winded way of quoting the mean and its uncertainty. A neater way is to write it as

$$v = (341.8 \pm 0.9) \text{ m/s}.$$

The above method of quoting the value is quite acceptable. However, you will very often see the final value quoted in *scientific notation* (as discussed in Section 2.5.3), so 314.8 becomes 3.418×10^2 and 0.9 becomes 0.009×10^2. Putting the two numbers together, along with the unit of measurement, we get

$$v = (3.418 \pm 0.009) \times 10^2 \text{ m/s}.$$

To summarise, after making repeated measurements of a quantity, there are four steps to take in quoting the value of the quantity:

Step 1 Calculate the mean of the measured values.
Step 2 Calculate the uncertainty in the quantity, making clear the method used. Round the uncertainty to one significant figure (or two if the first figure is a '1').
Step 3 Quote the mean and uncertainty to the appropriate number of figures.
Step 4 Include the units of the quantity.

The uncertainty that we quoted in our example was 0.9 m/s, which might suggest that all our measurements should lie between (341.8 − 0.9) m/s and (341.8 + 0.9) m/s,

---

[7] We consider this in more detail in Chapter 5.

[8] If the first figure in the uncertainty is a '1', such as in 1.46, it is quite common to give the uncertainty to two significant figures, i.e. in this case it would be ±1.5. To round 1.46 to 1 (i.e. to round to one significant figure) would be to suggest that the uncertainty is about 30% less than that calculated (i.e. 1 is about 30% less than 1.46). We will adopt this convention consistently in this book.

i.e. between 340.9 m/s and 342.7 m/s. Looking at the original data, we see that this is *not* the case. Of the eight measurements shown in Table 4.4, only *five* lie between 340.9 m/s and 342.7 m/s. Here we come to an important point: the uncertainty can be used to quantify the range in which the true value is likely to lie. Several of the individual values that contribute to the calculation of the uncertainty will always lie outside that range.

Exercise D

Below are statements of best estimates of quantities and their associated uncertainties. There is a mistake or omission in each of the statements. State the mistake that has been made and, where possible, correct it.

   (i)   The spring constant = (12.5731 ± 0.62841) N/m
  (ii)   The density of copper = (8825 ± 500) kg/m^3
 (iii)   The velocity of the sound wave in water = $(1.48 \pm 0.05) \times 10^3$
  (iv)   The unknown capacitance = (1.1 ± 0.001) μF
   (v)   The wavelength of the light = 6.0×10^{-7} m ± 1
  (vi)   The power produced by a solar cell array = $(9.51 \pm 0.5) \times 10^2$ W
 (vii)   The pH of the solution = 7.5 ± 0.07

4.3.3   Absolute, Fractional and Percentage Uncertainties

In most situations, it is advisable to express the uncertainty in the same units as the quantity being measured. This is referred to as the *absolute* uncertainty in the quantity. In some cases, however, you may be required to state the ratio (uncertainty in the quantity)/(quantity). This ratio is referred to as the *fractional* uncertainty in the quantity. For example, if the speed of an aircraft is (195 ± 5) m/s, then the fractional uncertainty in the speed is

$$\frac{5 \text{ m/s}}{195 \text{ m/s}} = 0.026.$$

Note that, as it is the ratio of two quantities that have the same units, the fractional uncertainty itself has *no* units.

The *percentage* uncertainty expresses the uncertainty as a percentage of the original quantity. It is found by multiplying the fractional uncertainty by 100%. So, in this example, the percentage uncertainty in the speed of the aircraft is

$$0.026 \times 100\% = 2.6\%.$$

As with absolute uncertainty, it is seldom necessary to quote fractional or percentage uncertainties to more than one significant figure. In the example above, we can reasonably round the percentage uncertainty to 3%.

Exercise E

Convert the following absolute uncertainties to fractional and percentage uncertainties:

(i) $I = (2.0 \pm 0.2)\,\text{A}$

(ii) $r = (1.56 \pm 0.07)\,\text{m}$

(iii) $t = (1.2 \pm 0.3) \times 10^6\,\text{s}$

(iv) $m = (5.6 \pm 0.6) \times 10^2\,\text{kg}$

(v) $T = (77 \pm 1)\,\text{K}$

(vi) $P = (9.5 \pm 0.5) \times 10^2\,\text{W}$

4.3.4   Accuracy and Precision

In everyday language, accuracy and precision signify much the same thing. In science and engineering, however, they have come to have different meanings. The difference can be explained if we first acknowledge that when we make a measurement of a quantity we are attempting to find an estimate of the true value of that quantity. How many measurements do we need to make before we find the exact or *true* value of a quantity? The answer is that the true value can never be known, but by gathering more data, then finding the mean of the values, we hope to get a better and better *estimate* of the true value. If our estimate, based on taking the mean of our data, is close to the true value, then we say that the measurements are *accurate*. But how do we know whether our mean is close to the true value?

Although the true value can never be known, it should be possible for workers in various laboratories endeavouring to measure the same quantity (for example the thermal conductivity of a new alloy) to agree that the true value lies within certain limits.[9] By better experimental methods, improved instruments or repetition of the measurements many times, it is possible to set about narrowing the limits. 'Best' values and uncertainties obtained for a quantity may be compared with those of other scientists and engineers, with a view to establishing consistency between workers. If consistency is not found, the methods and materials used by the workers need to be examined.

When we speak of a measurement being *precise* we mean that the uncertainty in the value is small, but this does not imply that it is close to the true value. This may

---

[9] We will see in Chapter 5 that we can quote a probability that the true value of a quantity lies between certain limits.

seem odd and even contradictory. How *can* a measurement be precise but not give a value close to the true value? Let us consider a particular example:

Thermocouples are widely used to measure temperature and may be connected to a suitably adapted digital voltmeter so that a direct reading of temperature can be made. In an experiment, high-purity water was poured into an open container and a heating element was used to bring the water to boiling point. While the water was boiling, 10 measurements of the temperature of the water were made using a thermocouple thermometer. The values obtained are shown in Table 4.6.

The mean of the values in Table 4.6 is 92.49 °C. The range of values is 0.4 °C, so that the uncertainty is 0.4 °C/10 = 0.04 °C. Therefore, we quote the boiling point of the water as

$$\text{boiling point} = (92.49 \pm 0.04)°\text{C}.$$

This looks fine but for one thing: the boiling point of pure water (at sea level, where the measurements were made) should be close to 100 °C and all of our data are lower than this by more than 7 °C. Our value for the boiling point has good precision, which implies that all the measured values are clustered tightly around the mean. However, it appears that the mean is some way away from the true value. The value for the boiling point is precise, but it is *not* accurate. The higher priority must be given to establishing an accurate value for the quantity under investigation, as precision without accuracy can only be misleading.

In a repeat of the above experiment, an alcohol-in-glass thermometer replaced the thermocouple. The scale on the thermometer could be read to the nearest 0.5 °C. The data obtained are shown in Table 4.7.

The mean of this set of data is 100.2 °C and the range is 2 °C. It follows that the uncertainty in the boiling point of the water, measured using the alcohol thermometer, is 2 °C/10 = 0.2 °C. We write

$$\text{boiling point of water} = (100.2 \pm 0.2)°\text{C}.$$

Table 4.6 Ten repeat measurements of the boiling point of water made using a thermocouple thermometer

Temp. (°C)	92.4	92.6	92.6	92.3	92.4	92.7	92.4	92.4	92.5	92.6

Table 4.7 Ten values of the boiling point of water obtained using an alcohol-in-glass thermometer

Temp. (°C)	101.0	100.5	99.0	99.0	99.5	100.5	100.0	101.0	100.5	101.0

Table 4.8 Summary of the difference between accuracy and precision	
If a value obtained through experiment is:	then:
accurate	it is close to the true value but, unless specified, the uncertainty could be of any magnitude
precise	it has a small uncertainty, but we cannot conclude that it is close to the true value
both accurate **and** precise	it is close to true value and with a small uncertainty. We would like our values to fall into this category

The measurement of the boiling point of the water is much less precise using the glass thermometer compared to the thermocouple thermometer (an uncertainty of 0.2 °C compared to 0.04 °C). In terms of accuracy, however, the alcohol thermometer has proved (in this experiment) to be superior to that of the thermocouple thermometer. So why were the measured values for the temperature of the boiling water found using the thermocouple so much lower than 100 °C? In answering that we must introduce the idea of systematic error. Before we deal with that topic, Table 4.8 summarises what we have learned about the difference between precision and accuracy.

4.4   Systematic and Random Errors

Why did the values obtained using the thermocouple thermometer in Section 4.3.4 differ so greatly from the expected value of 100 °C? To answer this, we must look more closely at the types of errors that can occur in an experiment. Broadly speaking we can place errors into two categories, referred to as *systematic* and *random* errors. We will come to random errors shortly, but it is worth stating here that such errors are usually easier to deal with than systematic errors because they are more readily identified, and the uncertainties that arise from random errors are easier to quantify using some basic statistical reasoning.

Systematic errors tend to be more serious than random errors, as systematic errors are often difficult to detect, and it is possible to perform an experiment while being completely unaware of their existence. In our example in Section 4.3.4, in which we were attempting to measure the boiling point of water, we were suspicious of the measurements, not because the measured values showed wide variation, but because they were so far away from the expected value of 100 °C. Once systematic errors

Temp. (°C)	−7.5	−7.3	−6.9	−7.4	−7.4	−7.7	−7.6	−7.6	−7.3	−7.6

Table 4.9 Ten values of the melting point of water obtained using a thermocouple

have been uncovered they can, to an extent, be corrected for.[10] It is their detection that challenges an experimenter.

Two types of systematic error which can exist with measuring instruments are *offset* error and *gain* error.

4.4.1   Offset Error

Consider an experiment to find the melting point of water using the thermocouple thermometer mentioned in Section 4.3.4 (which indicated a temperature of ~92.5 °C when placed in boiling water). The same thermocouple was placed in a pure ice and pure water mixture and 10 measurements of temperature were made. Table 4.9 shows the data obtained.

The mean of the above values is −7.43 °C and the range is 0.8 °C. Using our method for calculating the uncertainty, i.e. uncertainty = (range)/n, we find that the melting point of water is (-7.43 ± 0.08) °C. There is a problem here: the melting point of pure water should be very close to 0.0 °C. What we have uncovered is an offset error with our temperature measuring system of about −7.5 °C. For whatever reason (for example, low battery, malfunctioning digital meter or an incompatible type of thermocouple attached to the meter), all measurements of temperature are too low by about 7.5 °C. An offset of this size may not be of much consequence if you are measuring the temperature of a furnace which has been set to 1500 °C. By contrast, if you were trying to establish the body temperature of a newborn baby, a systematic error of −7.5 °C would most certainly *not* be acceptable. It is the extreme nature of the error, and the fact that we have some knowledge of the values to expect for the boiling and freezing points of water, that alerted us to the error. Clues such as these are absent in many measurement situations.

4.4.2   Gain Error

Another cause of systematic error can be attributed to gain error in the measurement system. In contrast to the offset error, which remains fixed irrespective of the magnitude of the quantity being measured, the gain error *is* dependent on the magnitude of the quantity. The effect of a gain error is best illustrated by example.

---

[10] No correction can completely eliminate a systematic error, as that would require us to know 'exactly' the size of the systematic error, and, in common with all errors, that is unknowable.

Five calibration masses (whose values are known to high accuracy) were placed in turn upon a top-loading electronic balance and the mass indicated by the balance was recorded. Table 4.10 shows the recorded values.

We can see that, as the mass placed on the balance increases, so the *difference* between measured and calibrated mass increases. If we use the symbol m_c to represent the calibrated mass and the symbol m_m to represent the measured mass, then Table 4.11 shows the calibration mass and the difference between m_c and m_m.

Figure 4.1 shows that the difference, $m_m - m_c$, in the two quantities increases in direct proportion to the magnitude of the mass on the balance. As these measurements have established the relationship between the calibrated mass and the measured mass (at least over the range 0 to 100 g), future measurements of mass using this balance can be corrected.

Table 4.10 Comparison between calibration masses and mass indicated by an electronic balance

Calibration mass (g)	0.00	20.00	40.00	60.00	80.00	100.00
Value indicated on balance (g)	0.00	20.26	40.65	60.98	81.20	101.52

Table 4.11 Difference between measured mass and calibration mass

m_c (g)	0.00	20.00	40.00	60.00	80.00	100.00
$(m_m - m_c)$ (g)	0.00	0.26	0.65	0.98	1.20	1.52

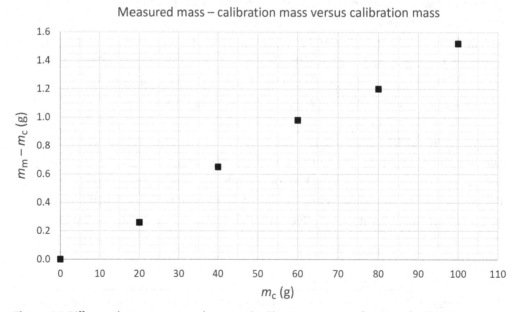

Figure 4.1 Difference between measured mass and calibration mass as a function of calibration mass.

4.4.3 Detecting and Dealing with Systematic Errors Caused by Instruments

When an instrument is used to measure a quantity, it is possible for that instrument to introduce an error into your values obtained through measurement. To identify offset and systematic gain errors, we must check the values indicated by the instrument against known standards across the range over which we wish to make measurements. If all the measured values differ from the standard values by the same amount, we have identified an offset error. As we saw in Section 4.4.2, a gain error depends on the magnitude of the quantity being measured. If the offset or gain errors can be estimated, the measured values can be adjusted to bring measured value of a quantity closer to the true value.[11]

4.4.4 Random Errors

Random errors produce scatter, such that values obtained through measurement are above and below the true value. The cause of the scatter could be the limitation in the scale of the instrument, so that on some occasions values are rounded up above the true value and at other times the values are rounded down below the true value.

In some circumstances a measurement is difficult to perform, and this is a cause of variation in the measured values. For example, in an optics experiment it is frequently difficult to find the position of 'best focus' of an image on a screen. The screen is moved back and forth in an effort to find the position where the image is sharpest. Although it may be possible to read the measuring scale to ±0.5 mm or better, the uncertainty in the image position could (in some cases) be many times this value.

Environmental factors can introduce random errors into measured values. Following are some examples.

- Electrical interference caused by the switching on and off of electrical equipment can affect sensitive voltage or current measurements.
- Vibrations caused by a passing motor vehicle (or even a passing person!) can unpredictably influence force measurements carried out using sensitive instruments such as electronic balances or electron microscopes.
- Variations in ambient temperature or humidity can affect measurements made using electronic equipment.
- In a water flow experiment, changes in mains water pressure can cause variations in water flow rates.

Although it may not be possible to identify all the sources of random error impacting on a measurement (and indeed several sources may be acting at the same

---

[11] We emphasise again that while offset and gain errors can be reduced, they cannot be eliminated.

time), we can use statistical techniques to give us an estimate of the uncertainty due to random errors and to allow us to calculate the effect of combining uncertainties. We will discuss the statistical approach to the addition of uncertainties in Chapter 5. There are situations in which we have insufficient data to justify a full statistical analysis, but still need to be able to estimate the effect of combining uncertainties. We will now consider methods by which uncertainties can be combined.

4.5   Combining Uncertainties

An experiment may require the determination of several quantities which are later to be entered into an equation. For example, we may measure the mass, m, of a body and its volume, V. The density of the body, ρ, can be calculated using the relationship

$$\rho = \frac{m}{V}. \tag{4.4}$$

How do uncertainties in m and V combine to give an uncertainty in ρ? We will now consider how to combine uncertainties.[12]

4.5.1   Combination of Uncertainties: Method I

This method requires only basic arithmetic. Each quantity in the equation is modified by an amount equal to the uncertainty in the quantity, to produce the largest value and smallest value resulting from application of the equation.

Example 1

Consider the calculation of the area of a wire of circular cross-section which has a diameter of (2.5 ± 0.1) mm. What is the area of cross-section and the uncertainty in that area?

The equation for the area, A, of a circle in terms of its diameter, d, is

$$A = \frac{\pi d^2}{4}. \tag{4.5}$$

Converting the diameter to metres and substituting into equation 4.5 gives

$$A = \frac{\pi \left(2.5 \times 10^{-3}\ \text{m}\right)^2}{4}$$
$$= 4.91 \times 10^{-6}\ \text{m}^2.$$

---

[12] This is sometimes referred to as propagation of uncertainties

This represents the best estimate of A, as it is based on the best value for the diameter.

To find the maximum cross-sectional area of the wire, we substitute the maximum value for the diameter, namely $(2.5 + 0.1)$ mm, i.e. 2.6 mm. The maximum area, A_{max}, is given by

$$A_{max} = \frac{\pi(2.6 \times 10^{-3}\text{ m})^2}{4}$$
$$= 5.31 \times 10^{-6}\text{m}^2.$$

To find the minimum area, A_{min}, we subtract the uncertainty in the diameter from the best value of the diameter, so that

$$A_{min} = \frac{\pi(2.4 \times 10^{-3}\text{ m})^2}{4}$$
$$= 4.52 \times 10^{-6}\text{m}^2.$$

The best estimate of the area is $4.91 \times 10^{-6}\text{ m}^2$, with an upper value of $5.31 \times 10^{-6}\text{ m}^2$ and lower value of $4.52 \times 10^{-6}\text{ m}^2$. We can quote the uncertainty, u_A, in the area by first subtracting A_{min} from A_{max} to give the *range* of A. This gives $0.79 \times 10^{-6}\text{ m}^2$. To find u_A, divide the range by 2:

$$u_A = \frac{A_{max} - A_{min}}{2} = 0.39 \times 10^{-6}\text{m}^2.$$

Now round this to $0.4 \times 10^{-6}\text{ m}^2$, so that $u_A = 0.4 \times 10^{-6}\text{ m}^2$. Finally, we quote the area as

$$A = (4.9 \pm 0.4) \times 10^{-6}\text{m}^2.$$

Example 2

In an electrical experiment, the current through a resistor was set at (2.5 ± 0.1) mA and the voltage across the resistor was found to be (5.5 ± 0.3) V. Calculate the resistance of the resistor using $R = V/I$ and the uncertainty in R, u_R:

$$R = \frac{V}{I} = \frac{5.5\text{ V}}{2.5 \times 10^{-3}\text{ A}} = 2.20 \times 10^3\ \Omega.$$

When dealing with the quotient of two quantities,[13] as in this example, the maximum value of the quotient will occur when the numerator is *increased* by

---

[13] The quotient of two numbers is the result of dividing one number by the other.

an amount equal to the uncertainty in the quantity appearing in the numerator and the denominator is *reduced* by an amount equal to the uncertainty in the quantity in the denominator. To find the minimum value of the quotient we use the minimum value for the numerator and the maximum value for the denominator.

Here the maximum value of the numerator is $(5.5 + 0.3)$ V and the minimum value of the denominator is $(2.5 - 0.1) \times 10^{-3}$ A. Therefore, the maximum resistance, R_{max}, is given by

$$R_{max} = \frac{V}{I} = \frac{5.8 \text{ V}}{2.4 \times 10^{-3} \text{ A}} = 2.42 \times 10^3 \text{ } \Omega.$$

R_{min} is found by using the minimum value of voltage and the maximum value of current. This gives $R_{min} = 2.00 \times 10^3$ Ω. To calculate the uncertainty in resistance, u_R, we have

$$u_R = \frac{R_{max} - R_{min}}{2} = 0.210 \times 10^3 \text{ } \Omega.$$

Rounding the uncertainty to one significant figure, we write the resistance as $R = (2.2 \pm 0.2) \times 10^3$ Ω.

Exercise F

(1) The mass of a block of wood is (3.15 ± 0.05) kg and its volume is $(6.35 \pm 0.02) \times 10^3$ cm^3.
   (i) Use equation 4.4 to determine the density of the wood. Express the density in kg/m^3.
   (ii) Determine the uncertainty in the density of the wood. Express the uncertainty in the density in kg/m^3.

(2) The radial acceleration, a, of a body rotating in a circle of radius r at constant speed v is given by

$$a = \frac{v^2}{r}.$$

If $v = (3.00 \pm 0.05)$ m/s and $r = (1.5 \pm 0.1)$ m, calculate a, the maximum and minimum values of a, and the uncertainty in a, u_a.

(3) The speed, v, of a transverse wave moving along a stretched string is given by

$$v = \left(\frac{T}{\mu}\right)^{\frac{1}{2}},$$

where T is the tension in the string and μ is the mass per unit length of the string. If $T = (25 \pm 2)\,\mathrm{N}$ and $\mu = (1.2 \pm 0.1) \times 10^{-2}\,\mathrm{kg/m}$, calculate v, the maximum and minimum values of v, and the uncertainty in v, u_v.

4.5.2 Combination of Uncertainties: Method II

Although the method just applied is quite reasonable for finding combined uncertainties, it is cumbersome, especially when the equation contains more than one quantity with uncertainty. We now introduce another technique for combining uncertainties, which represents an application of differential calculus. This may sound daunting and overly complicated, but in fact if we can differentiate functions such as sines, cosines and logarithms, we will encounter few difficulties in calculating uncertainties involving these functions.

Suppose V depends on the two variables, a and b. The mathematical way of writing this is

$$V = V(a, b).$$

We say that V is a *function* of a and b. An example of such a function would be

$$V = a^2 b.$$

If a changes by an amount δa and b changes by an amount δb, we can write the accompanying change in V, δV, as

$$\delta V = \frac{\partial V}{\partial a}\delta a + \frac{\partial V}{\partial b}\delta b. \tag{4.6}$$

$\partial V/\partial a$ is the partial derivative of V with respect to a, and $\partial V/\partial b$ is the partial derivative of V with respect to b. When finding a partial derivative, all quantities in the equation except for the one that is being differentiated with respect to, are taken to be constant.

As an example of partial differentiation, consider the function $V = a^2 b$. To find $\partial V/\partial a$, we treat b as a constant and a^2 differentiates to $2a$, to give $\partial V/\partial a = 2ab$. To find $\partial V/\partial b$, we treat a as a constant and b differentiates to 1. This gives $\partial V/\partial b = a^2$.

In order to use equation 4.6 in problems involving uncertainty, we replace the quantities δV, δa and δb by the uncertainties u_V, u_a and u_b, respectively, so that the equation is rewritten

$$u_V = \left|\frac{\partial V}{\partial a}\right| u_a + \left|\frac{\partial V}{\partial b}\right| u_b. \tag{4.7}$$

$|\partial V/\partial a|$ means that we take the magnitude of the partial derivative; that is, we ignore any minus sign that may occur once we have differentiated. The consequence of not ignoring the minus sign is that cancellation of the terms on the right-hand side of equation 4.7 could occur, leading unrealistically to an uncertainty of zero (or close to zero).

Note that:

- We partially differentiate the function with respect to each quantity that possesses uncertainty.
- Quantities with no uncertainty are regarded as constants.
- Equation 4.7 can be extended to any number of quantities. For example, if $V = V(a,b,c,)$ then $u_V = |\partial V/\partial a| u_a + |\partial V/\partial b| u_b + |\partial V/\partial c| u_c$.

Exercise G

(1) If $s = \frac{1}{2}at^2$, determine $\partial s/\partial a$ and $\partial s/\partial t$.
(2) If $P = I^2 R$, determine $\partial P/\partial I$ and $\partial P/\partial R$.
(3) If $n = \sin i/\sin r$, determine $\partial n/\partial i$ and $\partial n/\partial r$.
(4) If $v = (T/\mu)^{1/2}$, determine $\partial v/\partial T$ and $\partial v/\partial \mu$.
(5) If $f = uv/(u+v)$, determine $\partial f/\partial u$ and $\partial f/\partial v$.
(6) If $R = \rho l/A$, determine $\partial R/\partial \rho$, $\partial R/\partial l$ and $\partial R/\partial A$.

4.5.3 Combining Uncertainties: Sums, Differences, Products and Quotients

Equation 4.7 is applicable to any equation that you are likely to encounter. However, there are situations, such as taking the product of two quantities, that are so common that it is worth applying equation 4.7 to determine the specific relationship that combines the uncertainties in the quantities.

Sum: If $V = a + b$, and uncertainties in a and b are u_a and u_b, respectively, we can use equation 4.7 to find the uncertainty in V. We have

$$u_V = \left|\frac{\partial V}{\partial a}\right| u_a + \left|\frac{\partial V}{\partial b}\right| u_b. \tag{4.8}$$

Now $|\partial V/\partial a| = 1$ and $|\partial V/\partial b| = 1$, so that

$$\boxed{u_V = u_a + u_b.}$$

We see that the uncertainty in V is given by the sum of the uncertainties in a and b.

Difference: If $V = a - b$ then once again $|\partial V/\partial a| = 1$ and $|\partial V/\partial b| = 1$, so that

$$u_V = u_a + u_b,$$

i.e. when V is the difference between two quantities, the uncertainty in V is the *sum* of the uncertainties in a and b.

Product: If $V = ab$, then $|\partial V/\partial a| = b$ and $|\partial V/\partial b| = a$, so that using equation 4.7 we get $u_V = bu_a + au_b$. If we divide both sides of this equation by ab we get $u_V/(ab) = bu_a/(ab) + au_b/(ab)$, so that

$$\frac{u_V}{V} = \frac{u_a}{a} + \frac{u_b}{b}.$$

The uncertainty in the quantity divided by the quantity itself, in this case u_V/V, is the *fractional* uncertainty in V and is equal to the fractional uncertainty in a added to the fractional uncertainty in b.

Quotient: If $V = a/b$, then $|\partial V/\partial a| = 1/b$ and $|\partial V/\partial b| = a/b^2$, so that equation 4.7 becomes $u_V = u_a/b + au_b/b^2$. If both sides are divided by a/b, then we find that

$$\frac{u_V}{V} = \frac{u_a}{a} + \frac{u_b}{b}.$$

Example 3

The temperature of $(3.0 \pm 0.2) \times 10^2$ g of water is raised by $(5.5 \pm 0.5)\,°\text{C}$ by a heating element placed in the water. Calculate the amount of heat transferred to the water to cause this temperature rise. Also calculate the uncertainty in the amount of heat transferred to the water.

The equation relating heat input, Q, to temperature rise, θ, is

$$Q = mc\theta, \tag{4.9}$$

where m is the mass of the water and c is its specific heat capacity. We can find a value for c by referring to an established data source such as Kaye and Laby (1995). This source gives c as $4186\,\text{J/(kg\,K)}$ at $15\,°\text{C}$. Assuming that the uncertainty in c is small enough to be ignored, we can write

$$u_Q = \left|\frac{\partial Q}{\partial m}\right|u_m + \left|\frac{\partial Q}{\partial \theta}\right|u_\theta. \tag{4.10}$$

Now $|\partial Q/\partial m| = c\theta$ and $|\partial Q/\partial \theta| = mc$, so that equation 4.10 becomes[14]

---

[14] It is good practice not to round values in intermediate calculations, as this could lead to rounding errors accumulating and affecting the final value of a quantity and its uncertainty. It is better for rounding to occur once all calculations have been completed.

$$u_Q = c\theta u_m + mc u_\theta$$

$$= 4186 \text{ J/(kg K)} \times 5.5 \text{ K} \times 0.02 \text{ kg} + 0.3 \text{ kg} \times 4186 \text{ J/(kg K)} \times 0.5 \text{ K}$$

$$= 460.5 \text{ J} + 627.9 \text{ J} = 1088.4 \text{ J}.$$

Now we use equation 4.9 to calculate the heat, Q:

$$Q = 0.3 \text{ kg} \times 4186 \text{ J/(kg K)} \times 5.5 \text{ K} = 6907 \text{ J}.$$

We can now quote the energy transferred as $Q = (6.9 \pm 1.1) \times 10^3 \text{ J}$.

Example 4

In Section 4.5.1 the uncertainty in the cross-sectional area of a wire was calculated, given that the diameter of the wire was (2.5 ± 0.1) mm. The area, A, was found to be $(4.9 \pm 0.4) \times 10^{-6} \text{ m}^2$. Use the method given in this section to calculate the uncertainty in A.

We have that $A = \pi d^2/4$. u_A is the uncertainty in A, found by writing equation 4.7 as $u_A = |\partial A/\partial d| u_d$. Now, $|\partial A/\partial d| = 2\pi d/4 = \pi d/2$, so that

$$u_A = \frac{\pi d u_d}{2} = \frac{\pi \times 2.5 \times 10^{-3} \text{ m} \times 0.1 \times 10^{-3} \text{ m}}{2}$$
$$= 3.9 \times 10^{-7} \text{m}^2.$$

Rounding this to one significant figure gives $4 \times 10^{-7} \text{ m}^2$, which is equal to the value found for the uncertainty in Section 4.5.1.

Exercise H

In the following exercises, express your answers to the correct number of significant figures.

(1) The magnetic field, B, at the centre of a solenoid with n turns per metre, through which a current I passes, is given by $B = \mu_o n I$, where μ_o is the permittivity of free space $(\mu_o = 4\pi \times 10^{-7} \text{ H/m})$. If $n = (400 \pm 5) \text{ m}^{-1}$ and $I = (3.2 \pm 0.2) \text{ A}$, calculate B and the uncertainty in B, u_B.

(2) The current density, j, through a conductor of cross-sectional area S is given by $j = I/S$. If $I = (2.6 \pm 0.2) \text{ A}$ and $S = (2.5 \pm 0.1) \times 10^{-6} \text{ m}^2$, calculate j and the uncertainty in j, u_j.

(3) The focal length of a lens, f, is related to object distance, u, and image distance, v, by $1/f = 1/u + 1/v$. If $u = (30.0 \pm 0.1)\,\text{cm}$ and $v = (20.0 \pm 0.5)\,\text{cm}$, calculate f and the uncertainty in f, u_f.

(4) The number of active nuclei, N, that remain in a sample of radioactive material after a time t is given by $N = N_0 \exp(-\lambda t)$, where N_0 is the number of active nuclei when $t = 0$ and λ is the decay constant. If $N_0 = (6.0 \pm 0.4) \times 10^6$ and $t = (1.28 \pm 0.02) \times 10^3\,\text{s}$, calculate N and the uncertainty in N, u_N, given that $\lambda = 5.2 \times 10^{-3}\,\text{s}^{-1}$.

4.6 Selection and Rejection of Data

Data selection and rejection is a sensitive subject and one that can bring out strong feelings amongst experimenters. Some would argue that 'all data are equal' and that there are no circumstances in which the rejection of data can be condoned. At the other extreme there are those that 'know' that a set of data is spurious and reject it without a second thought. It is possible to feel sympathy for both views: if you have spent hours setting up an experiment and have confidence in the technique you are using, why *would* you reject any data you have collected? On the other hand, if you have operated the equipment many times and are familiar with the sort of numbers to expect from an experiment, it is not surprising that a set of data differing considerably from the rest is written off as 'spurious' and forgotten about.

There are statistical tests that can be applied which will assist in selecting data for rejection. However, the application of such tests, especially if done automatically by computer, can throw out data before anyone has had the opportunity to ask, 'are the data *truly* spurious or is there something going on that I should investigate?'

I consider that all data should be recorded as the experiment proceeds and that human (or computer) intervention that would have the effect of filtering the data should be avoided. This is not to say that all the data gathered should be used in further analysis. For example, if your concentration lapsed while you were hand-timing an event, it would be quite legitimate to neglect that timing in later calculations.

Confidence in an experiment and values obtained from that experiment really comes when you are satisfied that the measurements are repeatable. If you *do* have a suspect value, but you can see no reason to reject it, then good advice is to repeat the experiment. This may not always be possible. The experiment may consist of testing something to destruction. For example, in an experiment to study the relationship between stress applied to a wire and the strain[15] produced in that wire, you may be

---

[15] Stress is proportional to the force applied to the ends of the wire. Strain is proportional to the change in length of the wire that occurs as a consequence of the application of stress.

required to stretch the wire until it breaks. You can take another wire that is 'identical' to the first and repeat the experiment, but by choosing a new wire a very important element of the experiment has changed.

4.7   Comment

Errors in data are a part of life for experimenters in science and engineering. In this chapter we have discussed types of errors that can occur during an experiment and methods by which uncertainties arising from such errors can be combined. Although the approach we adopted when combining uncertainties is sufficient for many situations, it is possible that we are not getting the most from our data. In the next chapter we will study the distribution that provides a good description of variability for most data that we will gather: the normal distribution. This will lead us to another method of combining uncertainties.

Problems

4.1. Eight measurements of the resistance of a mercury sample were made at 300 K. The values obtained are given in Table 4.12.
   (i)   What is the range of the values in Table 4.12?
   (ii) Calculate the best estimate of the resistance and use the range to estimate the uncertainty in the best estimate.
   (iii) Express the best estimate and its uncertainty to the appropriate number of significant figures.

4.2. The ideal gas equation relates the pressure, P, volume, V, and temperature, T, for any gas at low pressure. The equation is $PV = nRT$, where R is a constant with value 8.314 J/(K mol) and n is the number of moles of gas present. If $P = (0.6 \pm 0.1) \times 10^5$ Pa, $V = (22 \pm 2) \times 10^{-3}$ m³ and $T = (325 \pm 5)$ K, calculate n and the uncertainty in n (assume that the uncertainty in R is negligible).

4.3. As part of a study on the quality of river water, the following characteristics of the water were measured: flow rate, pH, temperature, electrical conductivity and lead content. Table 4.13 shows values from repeat measurements of each quantity.

Table 4.12 Values from repeat measurements of the resistance of a sample of mercury

Resistance ($\times 10^{-2}$ Ω)	9.5	9.3	9.9	9.9	9.1	8.9	9.6	9.3

Table 4.13 River water data

Flow rate (m/s)	pH	Temperature (°C)	Electrical conductivity (μS/cm)	Lead content (ppb)[a]
0.45	6.9	10.0	906	37
0.48	6.8	11.0	1 105	54
0.46	7.3	11.0	1 004	34
0.39	7.2	9.5	998	45
0.41	7.2	10.0	885	68
0.41	6.9	10.5	780	56
0.46	7.0	11.5	885	70

[a] ppb stands for 'parts per billion'.

    (i) For each quantity shown in Table 4.13, calculate the best estimate of the quantity and the uncertainty in the best estimate of the quantity, using the methods discussed in Sections 4.3 and 4.3.1.

    (ii) Which quantity exhibits the largest *fractional* uncertainty?

4.4. The period of the motion, T, of a body of mass M on the end of a spring of spring constant k is given by $T = 2\pi\sqrt{M/k}$. Given that $M = (210 \pm 5)\,\text{g}$ and $T = (1.1 \pm 0.1)\,\text{s}$:

    (i) calculate k

    (ii) use the method described in Section 4.5.1 to find the uncertainty in k

    (iii) repeat the calculation of uncertainty using the method described in Section 4.5.2.

4.5. The gain, G, of an amplifier in decibels may be written $G = 20\log_{10}(V_{out}/V_{in})$, where V_{out} and V_{in} are the output and input voltages of the amplifier respectively. A student finds the output voltage of an amplifier to be $(122 \pm 3)\,\text{mV}$ and the input voltage to be $(33 \pm 2)\,\text{mV}$. Calculate the gain of the amplifier and the uncertainty in the gain. Hint: if you are going to use the method for combining uncertainties discussed in Section 4.5.2, then it is useful to know that if $y = \log_{10}(u)$, then $\partial y/\partial u = 0.4343/u$.

4.6. Given that $z = ae^b$, find $\partial z/\partial a$ and $\partial z/\partial b$. Using this information, and given that $a = (4.6 \pm 0.2)$ and $b = (10.2 \pm 0.1)$, find z and the uncertainty in z.

4.7. Table 4.14 shows the times taken for a reaction to occur between sodium thiosulfate and hydrochloric acid. Included in the table are the uncertainties in the values of reaction time.

There is a linear relationship between 1/(reaction time) and the concentration. Attach two columns to Table 4.14, one showing 1/(reaction time) and the other the uncertainty in this quantity.

Table 4.14 Time for reaction between $Na_2S_2O_3$ and hydrochloric acid as a function of concentration of $Na_2S_2O_3$

Concentration (mol)	Reaction time, t (s)
0.20	25 ± 2
0.16	30 ± 2
0.14	32 ± 2
0.12	40 ± 3
0.08	56 ± 3
0.02	116 ± 3

4.8. The coefficient of static friction, μ_s, for two surfaces in contact is related to a critical angle, θ_c, to which the surfaces are tilted, by the equation $\mu_s = \tan \theta_c$. Given that $\theta_c = (44 \pm 1)°$, determine μ_s and the uncertainty in μ_s.

4.9. The pressure difference, p, between two points in a flowing fluid is related to the density of the fluid, ρ, and the speed of the fluid, v, by the equation $p = \frac{1}{2}\rho v^2$. Given that $\rho = (1.02 \pm 0.02) \times 10^3 \, kg/m^3$ and $v = (0.84 \pm 0.03) \, m/s$, calculate p and the uncertainty in p.

5 Statistical Approach to Variability in Measurements

5.1 Overview: Estimating Uncertainties with the Aid of Statistics

No matter how much care we take during an experiment, or how sophisticated the equipment we use, values obtained through measurement are influenced by errors. We can think of errors as acting to conceal the true value of the quantity sought through experiment. Random errors cause values obtained through measurement to occur above and below the true value.

In this chapter we will consider statistically based methods for dealing with variability in experimental data such as that caused by random errors. As statistics can be described as the science of assembling, organising and interpreting numerical data, it is an ideal tool for assisting in the analysis of experimental data.

The approach we will adopt in this chapter will be to estimate the uncertainty in a quantity that results from the impact of random errors on the spread of data obtained through repeated measurements of the quantity, and will not be concerned with assessing the uncertainties in individual measurements. This approach is valid when we have made sufficient measurements (say more than five) which reasonably describe the spread in data due to random errors.

We will not derive the formulae appearing in this chapter, but will indicate through examples their credibility and applicability.[1]

5.2 Variance and Standard Deviation of Repeated Measurements

The mean of values obtained by repeated measurements is regarded as the best estimate of the true value of the quantity being measured. How may these values be used to provide a single number which represents the uncertainty in that best estimate? To answer that question, we will begin by looking at a typical set of data.

---

[1] For more details on the statistics of error and uncertainty, see Berendsen (2011).

Table 5.1 Ten values of the time for a body to slide down an inclined plane

Time (s)	0.64	0.64	0.59	0.58	0.70	0.61	0.68	0.55	0.57	0.63

Table 5.2 Values, deviations from mean and (deviation from mean)^2

x_i (s)	$d_i = x_i - \bar{x}$ (s)	$d_i^2 = (x_i - \bar{x})^2$ (s^2)
0.64	0.021	0.000441
0.64	0.021	0.000441
0.59	−0.029	0.000841
0.58	−0.039	0.001521
0.70	0.081	0.006561
0.61	−0.009	0.000081
0.68	0.061	0.003721
0.55	−0.069	0.004761
0.57	−0.049	0.002401
0.63	0.011	0.000121
	$\sum (x_i - \bar{x}) = 0$	$\sum (x_i - \bar{x})^2 = 0.02089$ s^2

In a study of the motion of a body as it slides down an inclined plane, the time for the body to slide 80 cm was measured using a hand-held stopwatch. Table 5.1 shows values from 10 repeat timings of the motion.

The mean of the numbers in Table 5.1 is 0.619 s. To estimate the spread or variability in the data, we begin by drawing up a table which indicates how far each value is from the mean. The difference between the ith value and the mean is called the *deviation* and is represented by the symbol d_i. Table 5.2 shows the deviation from the mean, and the square of the deviation from the mean, of the data in Table 5.1.

At first sight it might seem reasonable to take the mean of the deviations shown in the second column of Table 5.2 as a measure of the variability of the data, but there is a difficulty with this: as we can see from the bottom row of the second column, the sum of the deviations is zero,[2] so that the mean deviation is also zero. We need to look elsewhere for a number which represents the spread of the data.

What *is* most often taken as a measure of the variability of the values is the mean of the sum of the *squares* of the deviations. This is referred to as the

---

[2] In fact, this is just another way of defining the mean.

variance of the values, and is usually represented by the symbol σ^2. The variance is defined as

$$\sigma^2 = \frac{\sum d_i^2}{n},$$

so that

$$\sigma^2 = \frac{\sum (x_i - \bar{x})^2}{n}, \tag{5.1}$$

where n is the number of repeat measurements.

The third column in Table 5.2 shows the squares of the deviations and their sum, which is $0.02089\,\mathrm{s}^2$. It follows that

$$\sigma^2 = \frac{0.02089\,\mathrm{s}^2}{10} = 0.002089\,\mathrm{s}^2.$$

Notice that the unit of variance is the *square* of the unit in which the original measurements were made.

Although variance is a fundamental measure of the spread of values, it is more common to use the *standard deviation* of the values to represent the spread. The standard deviation has the same units as those of the original measurements. The standard deviation is defined as (variance)$^{1/2}$, and is often represented by the symbol σ. Using equation 5.1 we get

$$\sigma = \left(\frac{\sum (x_i - \bar{x})^2}{n} \right)^{1/2}. \tag{5.2}$$

For the values in Table 5.1, the standard deviation is $(0.002089\,\mathrm{s}^2)^{1/2} = 0.04571\,\mathrm{s}$.

Statistical functions, such as that given by equation 5.2, can be found on many scientific calculators and on calculator applications available for smartphones.[3] These functions aid calculations considerably, especially when there are many values to process.

If we were to repeat the timing experiment considered in this section, but instead making 50 repeat measurements, how would the mean and standard deviation be affected? Table 5.3 contains 50 consecutive measurements of time. It is left as an exercise to show that the mean of the values in Table 5.3 is $0.6042\,\mathrm{s}$ and the standard deviation, calculated using equation 5.2, is $0.04364\,\mathrm{s}$.

Table 5.4 shows a comparison of the means and standard deviations for 10 and 50 measurements of the time for a body to slide down an inclined plane, and shows

---

[3] An example is Stats Calculator Free for Android smartphones (https://play.google.com/store/apps/details?id=me.nickpierson.StatsCalculator&hl=en).

Table 5.3 Fifty consecutive measurements of time for a body to slide down a plane

					Time (s)				
0.61	0.58	0.60	0.60	0.53	0.64	0.64	0.59	0.58	0.70
0.60	0.51	0.69	0.66	0.56	0.61	0.68	0.55	0.57	0.63
0.61	0.64	0.60	0.61	0.58	0.62	0.64	0.53	0.59	0.63
0.61	0.63	0.65	0.64	0.58	0.61	0.68	0.62	0.61	0.60
0.68	0.56	0.59	0.53	0.55	0.59	0.62	0.54	0.59	0.55

Table 5.4 Means and standard deviations compared for 10 and 50 repeat measurements of the same quantity

	Mean, $\bar{x}$ (s)	Standard deviation, σ (s)
10 measurements	0.6190	0.04571
50 measurements	0.6042	0.04364

that the standard deviation is little changed by taking 50 measurements rather than 10.

In general, the following statement holds for the standard deviation:

The standard deviation of values obtained by repeat measurements of a quantity remains almost constant, regardless of how many measurements are made.

Questions that naturally arise are: should we take the standard deviation as the uncertainty in the best estimate of the true value of a quantity and, if so, why make many repeat measurements if the standard deviation can be calculated using only a few?

Let us review the reason for making repeated measurements of a quantity during an experiment. We are trying to find the best estimate for the quantity, which we take to be the mean, but in addition we want the uncertainty in the best estimate. The standard deviation is representative of the spread of the *whole* data set and therefore should not be taken as the uncertainty in the best estimate.

We introduce a new term, the *standard error* of the mean, represented by the symbol $\sigma_{\bar{x}}$,[4] when speaking of the uncertainty in the best estimate of a quantity. By reference to a specific example, we will explain the standard error and how it is related to the standard deviation of repeated measurements.

---

[4] In Section 5.4.2 we will have cause to use the symbol $s_{\bar{x}}$ in preference to $\sigma_{\bar{x}}$ to represent the standard error of the mean.

5.3   Uncertainty in the Best Estimate of the True Value Obtained through Repeat Measurements: Standard Error of the Mean

In a fluid flow experiment, 10 measurements were made of the volume of water flowing through the apparatus in one minute. Table 5.5 shows the data gathered.

The best estimate of the true value of the volume is the mean of the values in Table 5.5, which is 41.7 ml. The standard deviation of the values, calculated using equation 5.2, is 3.29 ml.

The experiment was performed eight times in all, with each experiment consisting of 10 repeat measurements of the volume of liquid. Table 5.6 shows the mean and standard deviation for each of the eight sets of measurements.

By comparing the means in Table 5.6 with the 10 values in Table 5.5, we see that the variability in the means is considerably less than the variability in the original data. We might expect this, as the reason for calculating the mean is to 'average out' the scatter due to the random errors which influence individual measurements.

The mean of the means in Table 5.6 is 41.1 ml, and the standard deviation of the means is 0.893 ml. The standard deviation of the means is the *standard error of the mean*, $\sigma_{\bar{x}}$.

We can see that the 'mean of the means' differs little from the individual means in Table 5.6. However, the standard error of the mean is less than the standard deviation in the original data by a factor of approximately 3. As the number of repeat measurements in each experiment is equal to 10, we can see that, for this example at least, we could write the relationship between $\sigma_{\bar{x}}$ and σ as

$$\sigma_{\bar{x}} \approx \frac{\sigma}{\sqrt{n}}, \tag{5.3}$$

where n is the number of repeat measurements.

Table 5.5 Values of the volume of water collected in a fluid flow experiment

Volume of liquid collected (ml)									
33	45	43	42	45	42	41	44	40	42

Table 5.6 Means and standard deviations for eight sets of values (each set consists of 10 values)

Mean (ml)	41.0	41.7	40.4	41.5	41.7	40.4	42.5	39.5
Standard deviation (ml)	3.13	3.29	3.07	3.11	3.20	2.94	2.73	3.20

A mathematical proof[5] would show that the '$\approx$' symbol in equation 5.3 can be replaced by the '=' symbol. We can write

$$\sigma_{\bar{x}} = \frac{\sigma}{\sqrt{n}}. \tag{5.4}$$

When we make repeat measurements of a quantity, it is $\sigma_{\bar{x}}$ that we take as the uncertainty in the best estimate of the true value.

We can now return to the data in Table 5.4 and use equation 5.4 to obtain the uncertainty in the best estimate of the time when 10 and 50 measurements are made.

The uncertainty in the best estimate based on the 10 measurements of time is

$$\sigma_{\bar{x}} = \frac{\sigma}{\sqrt{n}} = \frac{0.04571 \text{ s}}{\sqrt{10}} = 0.0144 \text{ s}.$$

The best estimate of the true time based on 10 repeat measurements is[6]

$$(0.619 \pm 0.014) \text{ s}.$$

For the set of 50 measured values we have

$$\sigma_{\bar{x}} = \frac{\sigma}{\sqrt{n}} = \frac{0.04364 \text{ s}}{\sqrt{50}} = 0.0062 \text{ s}.$$

The best estimate of the true time based on 50 repeat measurements is

$$(0.604 \pm 0.006) \text{ s}.$$

We see that it *is* worthwhile to make more measurements of a quantity if we want to reduce the uncertainty in the best estimate of the true value of that quantity, where that uncertainty is due to random errors. However, to reduce the uncertainty by a factor of 2, equation 5.4 indicates that we must increase the number of measurements by a factor of 4.

To summarise:

(i)   The best estimate of the true value of a quantity is found by taking the mean of values obtained through repeated measurements of that quantity.
(ii) The standard deviation is a measure of the spread of the values and is quite insensitive to how many measurements are made.
(iii) The standard error of the mean is taken as the uncertainty in the best estimate and this *does* decrease as the number of repeat measurements increases.

---

[5] We will not derive this equation. For a derivation, see Taylor (1997), chapter 5.
[6] As in Chapter 4, we will continue to apply the convention that uncertainties should be quoted to one significant figure (unless the first figure is a '1').

Exercise A

(1) Calculate the mean, standard deviation and standard error of the mean of the numbers in Table 5.7.

(2) A steel ball is dropped from a height of 400 mm onto a metal plate. The rebound heights of the ball are shown in Table 5.8.

Using the values in Table 5.8, calculate the mean, variance, standard deviation and standard error of the mean rebound height. Give the best estimate of the rebound height and the uncertainty in the best estimate, to an appropriate number of significant figures.

Table 5.7 Eight numbers

25.2	23.3	22.7	27.0	26.8	24.0	26.6	22.9

Table 5.8 Sixty values of the rebound height of a ball

Rebound height (mm)					
176	173	174	176	175	173
176	176	176	178	176	171
178	176	174	176	173	174
177	176	177	176	175	176
175	176	178	176	172	177
179	169	176	178	173	177
178	178	176	176	176	175
178	170	178	176	177	178
178	179	176	180	178	178
177	173	177	176	180	173

5.4 Displaying the Values Obtained from Repeated Measurements: The Histogram

A revealing way to display data from many repeat measurements of a quantity is to draw a chart, referred to as a *histogram*. The range of the data is divided into a number of equal intervals and the number of values (referred to as the *frequency*) that fall into each interval[7] is plotted vertically with the intervals plotted horizontally.

---

[7] An interval in a histogram is sometimes referred to as a 'bin'.

Table 5.9 Ninety values of the temperature of a water bath (°C)

52.9	44.8	51.6	43.0	45.0	46.5	37.2	48.9	49.4
52.5	45.9	49.0	44.7	38.5	50.5	50.8	49.8	38.1
44.0	48.7	52.3	44.2	47.2	42.9	38.8	41.1	50.3
42.8	43.3	47.1	49.1	45.4	47.2	45.9	42.0	50.7
48.1	48.3	52.9	46.0	47.4	48.6	46.5	44.2	36.5
50.7	43.7	42.8	42.9	35.6	46.9	51.3	46.1	47.5
47.8	47.8	52.4	39.7	35.6	46.8	38.3	50.5	42.6
55.4	45.4	35.9	47.4	40.0	49.9	37.2	42.5	40.1
46.9	45.7	52.1	40.5	41.1	50.0	39.3	42.1	48.4
40.7	41.1	44.4	49.6	47.9	41.0	46.4	46.5	39.0

Table 5.9 contains 90 values of the temperature of a water bath, to be displayed in the form of a histogram.

When deciding how many intervals should appear in a histogram there are no hard and fast rules to follow. One method of constructing a histogram is to make the number of intervals approximately equal to the square root of the number of values in the data set. The width of the intervals appearing along the horizontal axis is then made equal to the range divided by the number of intervals. In some cases (as in this example), the width of the interval is rounded in order to make the counting of the number of values in each interval easier (even if this requires the total number of intervals to increase or decrease slightly).[8] Rounding makes the horizontal axis of the histogram look neater.[9]

The first column in Table 5.10 indicates the steps by which a histogram can be constructed. The second column shows those steps applied to the data in Table 5.9. Figure 5.1 shows the data from Table 5.9 presented in the form of a histogram.

The method of constructing a histogram is now applied to the timing data shown in Table 5.3. The histogram is shown in Figure 5.2.

The histogram in Figure 5.2 has the following features:

(i)   The interval 0.58 s to 0.62 s has the largest frequency. This interval includes the mean and indicates that a large proportion of values lie close to the mean.

(ii) The spread or *distribution* of values is approximately symmetric about the interval containing the mean.

(iii) There are few values that lie far from the mean.

---

[8] The Excel spreadsheet contains a tool for creating a histogram. This will be discussed in Chapter 8.

[9] Histograms prepared automatically by some computer packages can produce 'ugly' intervals, requiring you to step in to change the default interval width.

Table 5.10 Step-by-step approach to plotting a histogram

Steps to take when plotting a histogram	Worked example based on data shown in Table 5.9	
1. Find the range, R, of the data (max. value − min. value).	$R = 55.4\,°C - 35.6\,°C = 19.8\,°C$	
2. Count total number of values in data set, N.	$N = 90$	
3. Calculate $\sqrt{N}$ and round to the nearest whole number. This is taken as the number of intervals.	$\sqrt{N} = 9.487$, rounds to 9	
4. Divide R by the number calculated in step 3 to give width of each interval.	width of interval = 2.2, rounds to 2	
5. Draw up table showing all the intervals covering the range and the number of data values (i.e. the frequency) in each interval. Make it clear into which interval a number on the borderline between two intervals (e.g. 45 in the worked example) should be placed. Note: an effect of rounding the interval width to 2 in this example is to increase the total number of intervals required from 9 to 11.	Interval (°C)	Frequency
	$35 \leq x < 37$	4
	$37 \leq x < 39$	6
	$39 \leq x < 41$	7
	$41 \leq x < 43$	12
	$43 \leq x < 45$	9
	$45 \leq x < 47$	15
	$47 \leq x < 49$	15
	$49 \leq x < 51$	13
	$51 \leq x < 53$	8
	$53 \leq x < 55$	0
	$55 \leq x < 57$	1

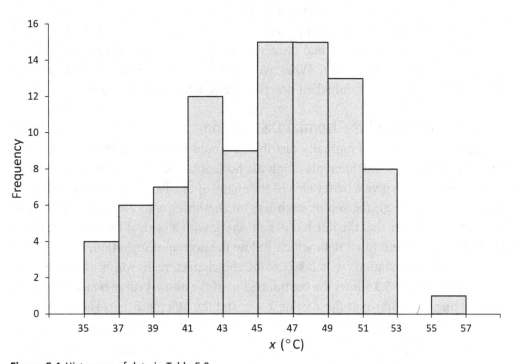

Figure 5.1 Histogram of data in Table 5.9.

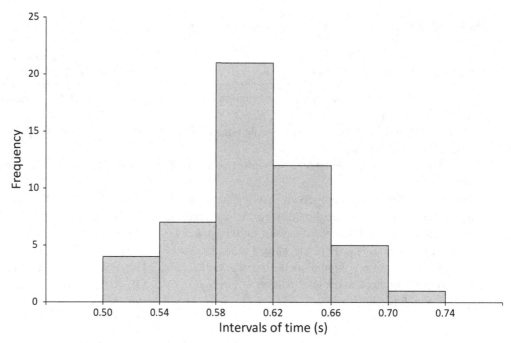

Figure 5.2 Histogram of data in Table 5.3.

Characteristics (i) to (iii) are displayed by data gathered in such a wide variety of experiments that a brief discussion of a statistical distribution that possesses these properties is in order.

One distribution of data which exhibits the above characteristics is referred to as the *normal* distribution,[10] and data from experiments are often referred to as being 'normally distributed'. The normal distribution can be derived from first principles, but we will not do that here. What we need to know are the features of the distribution which are useful when describing and analysing experimental data.

5.4.1   Properties of the Normal Distribution

A large data set which is normally distributed would produce a histogram like that shown in Figure 5.3. The intervals along the horizontal axis have been chosen to be very narrow and to give a better idea of the shape of the distribution a line is drawn which passes through the top of each bar on the histogram, all other lines being omitted. We can see that the line has a 'bell' shape with a peak at the mean, $\bar{x}$. These are features common to all data which follow the normal distribution.

The standard deviation, σ, is taken to be the characteristic width of the normal distribution. Figure 5.3 shows the central region of the normal curve bounded by two vertical lines, one drawn at the x value $\bar{x} - \sigma$ and the other at $\bar{x} + \sigma$. The area under

---

[10] Also referred to as the Gaussian distribution.

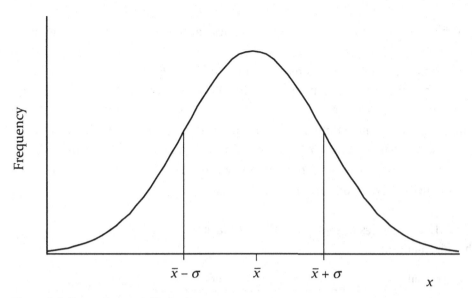

Figure 5.3 Shape of normal distribution.

the curve between the two lines is proportional to the number of values lying between $\bar{x} - \sigma$ and $\bar{x} + \sigma$. It can be shown that almost 70% of the total area under the whole curve lies within $\pm\sigma$ of the mean, indicating that ~70% of the values lie between $\bar{x} - \sigma$ and $\bar{x} + \sigma$. In addition, ~95% of values lie between $\bar{x} - 2\sigma$ and $\bar{x} + 2\sigma$.

Exercise B

In an optics experiment, 60 repeat measurements of the distance between a mirror and an image produced by the mirror were made. The values obtained are shown in Table 5.11.

Table 5.11 Sixty values of image distance

Image distance (mm)					
146	174	170	165	175	167
163	171	161	166	157	173
162	148	180	165	175	176
158	174	161	176	166	158
164	169	157	153	175	159
159	160	174	156	166	166
173	158	168	166	173	159
151	160	155	168	166	165
164	175	161	171	167	166
164	163	156	162	167	164

Table 5.12 presents the number of values falling into some of the given intervals.

(i) Complete Table 5.12 using the values in Table 5.11.

(ii) Draw up a histogram using Table 5.12. Do the values look to be normally distributed?

(iii) The mean and standard deviation of the values in Table 5.11 are $\bar{x} = 164.8$ mm and $\sigma = 7.3$ mm, respectively. How many values lie within $\pm\sigma$ of the mean and how many within $\pm 2\sigma$ of the mean? Is this consistent with the normal distribution?

Table 5.12 Frequencies of experimental values falling within given intervals

Interval, x (mm)	Frequency
$145 \leq x < 150$	2
$150 \leq x < 155$	2
$155 \leq x < 160$	11
$160 \leq x < 165$	
$165 \leq x < 170$	
$170 \leq x < 175$	
$175 \leq x < 180$	
$180 \leq x < 185$	

5.4.2   Population and Sample

Although we must do all we can to ensure that data gathered during an experiment are repeatable and have uncertainties reduced to as small a level as possible, this does not mean we can devote all our time to making measurements. Though important, data gathering is only one aspect of a well-planned and well-executed experiment. The totality of measurements that *could* be made is referred to as the *population* of measurements with mean, μ, and standard deviation, σ_{POP}, given by[11]

$$\mu = \frac{\sum_{i=1}^{i=\infty} x_i}{n} \tag{5.5}$$

$$\sigma_{POP} = \left(\frac{\sum_{i=1}^{i=\infty} (x_i - \mu)^2}{n} \right)^{\frac{1}{2}}, \tag{5.6}$$

---

[11] A number which is characteristic of the population of data is referred to as a *population parameter*.

where n is the totality of measurements that could be made (which, in principle, could approach infinity).

We can take the population mean, μ, to be the true value of the quantity being sought when making a measurement.[12] That this value can only be found as n approaches infinity makes it of little practical use. What we can do is to obtain values through a few repeat measurements which are a *sample* of all possible measurements, and estimate the population mean (and standard deviation) using those values. The estimate of the population mean is simply the mean, $\bar{x}$, of the values.

The symbol s is used to represent the estimate of the population standard deviation, and is calculated using an equation very similar to equation 5.2:[13]

$$\sigma_{POP} \approx s = \left(\frac{\sum (x_i - \bar{x})^2}{n - 1} \right)^{1/2}. \tag{5.7}$$

Equation 5.7 allows us to estimate the population standard deviation using a sample taken from the population. This contrasts with equation 5.2, which gives the standard deviation of the sample (Appendix 1 explains why $n - 1$, rather than n, appears in equation 5.7).

We want information concerning the population from which a sample is drawn; therefore equation 5.7 is preferred to equation 5.2 when calculating the standard deviation. In practice, however, so long as the number of repeated measurements is greater than 3, equations 5.2 and 5.7 will usually return the same number to one significant figure. This is good enough for most uncertainty calculations involving the standard deviation. The larger the value of n, the closer will be the numbers produced by equations 5.2 and 5.7.

We can calculate the standard error of the mean, $s_{\bar{x}}$, using s instead of σ, as

$$s_{\bar{x}} = \frac{s}{\sqrt{n}}. \tag{5.8}$$

Though in practice there is usually little difference between the standard error of the mean obtained from equations 5.4 and 5.8, we will generally favour equation 5.8, as this recognises that our sample size is finite.[14]

---

[12] This is valid if systematic errors are so small that they can be ignored. If systematic errors are not small, then no matter how many repeat measurements are made of a quantity, the effect of these errors remains, causing the population mean to differ from the true value.

[13] As usual, we assume that all summations, such as that appearing in equation 5.7, are taken between $i = 1$ and $i = n$.

[14] So long as n is 12 or less, equation 4.3, i.e. uncertainty = (range)/n, which we introduced in Section 4.3.1, provides a reasonable approximation of $s_{\bar{x}}$.

Exercise C

(1) Calculate $\bar{x}$, σ and s for the numbers in Table 5.13.
(2) Calculate the standard error of the mean, first using equation 5.4 and then equation 5.8. Give your answers to:
   (i) three significant figures
   (ii) one significant figure.

Table 5.13 Twelve numbers

86	70	75	78	90	59	78	82	81	70	70	69

5.4.3   Confidence Limits and Confidence Interval

The distribution of means of samples taken during an experiment follows a normal distribution in the same way as the 'raw' data. Figure 5.4 shows the distribution of means.

The major difference between the distribution of the raw data and the distribution of the means is in the width of the distribution. The width is given by the standard deviation, σ, in the case of the individual measurements, and by $\sigma_{\bar{x}} = \sigma/\sqrt{n}$ for the means, where n is the number of repeat measurements made.

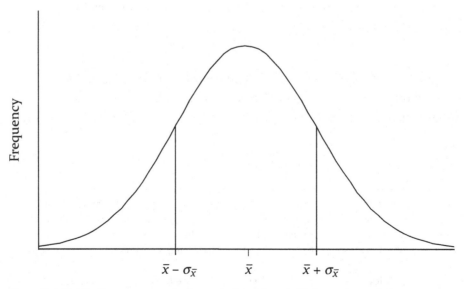

Figure 5.4 Distribution of means of values calculated from typical experimental data.

Table 5.14 Confidence intervals and the probability that the true value of a quantity lies within these intervals	
Confidence interval	Probability that true value lies within the confidence interval (%)
$\bar{x} - \sigma_{\bar{x}}$ to $\bar{x} + \sigma_{\bar{x}}$	68.3
$\bar{x} - 2\sigma_{\bar{x}}$ to $\bar{x} + 2\sigma_{\bar{x}}$	95.4
$\bar{x} - 3\sigma_{\bar{x}}$ to $\bar{x} + 3\sigma_{\bar{x}}$	99.7
$\bar{x} - 4\sigma_{\bar{x}}$ to $\bar{x} + 4\sigma_{\bar{x}}$	99.994

We saw in Section 5.4.1 that approximately 70% of data lie within $\pm\sigma$ of the mean and approximately 95% within $\pm 2\sigma$. When dealing with the standard error of the mean we can interpret $\pm\sigma_{\bar{x}}$ as follows. The true value has:

- approximately 70% chance of lying within $\pm\sigma_{\bar{x}}$ of the mean of the sample
- approximately 95% chance of lying within $\pm 2\sigma_{\bar{x}}$ of the mean of the sample.

We see that it is possible for the true value of a quantity to be outside the limits $\bar{x} - 2\sigma_{\bar{x}}$ to $\bar{x} + 2\sigma_{\bar{x}}$. However, that probability is low at 5%, or 1 chance in 20. We speak of a *confidence interval* bounded by *confidence limits* determined with the aid of the standard error of the mean. Table 5.14 shows a summary of confidence intervals and their associated probabilities.

Example 1

In an experiment to study the diffusion of carbon into iron, the mean value for the diffusivity, $\bar{D}$, of a number of repeat measurements was found to be $2.75 \times 10^{-7} \mathrm{cm^2/s}$, with a standard error of the mean, $\sigma_{\bar{D}}$, of $0.07 \times 10^{-7}\ \mathrm{cm^2/s}$. What is the 95% confidence interval for the true value of the diffusivity?

Using Table 5.14, the 95% confidence interval is $\bar{D} - 2\sigma_{\bar{D}}$ to $\bar{D} + 2\sigma_{\bar{D}}$, i.e. $(2.75 \times 10^{-7} - 2 \times 0.07 \times 10^{-7})\ \mathrm{cm^2/s}$ to $(2.75 \times 10^{-7} + 2 \times 0.07 \times 10^{-7})\ \mathrm{cm^2/s} = 2.61 \times 10^{-7}\mathrm{cm^2/s}$ to $2.89 \times 10^{-7}\mathrm{cm^2/s}$. This can be written neatly as $(2.75 \pm 0.14) \times 10^{-7}\mathrm{cm^2/s}$.

As for uncertainties, it is reasonable to give the number following the $\pm$ sign to one significant figure unless the first figure is a '1' (as in this example), in which case the number is given to two significant figures.

5.4.4 Replacing $\sigma_{\bar{x}}$ by $s_{\bar{x}}$

If we replace $\sigma_{\bar{x}}$ by $s_{\bar{x}}$ in Table 5.13, does the confidence interval stay the same? For example, is the probability 95.4% that the true value lies between the limits

$\bar{x} - 2s_{\bar{x}}$ and $\bar{x} + 2s_{\bar{x}}$? The strict answer is no. The reason is that the distribution of $s_{\bar{x}}$ is broader[15] than that of $\sigma_{\bar{x}}$ and increases in width as the number of repeat measurements decreases (for large n, say over 30, the distribution of $s_{\bar{x}}$ is very close to that of $\sigma_{\bar{x}}$). When the number of repeated measurements is small, the normal distribution tends to underestimate the size of the confidence interval, and the t distribution is preferred. Appendix 1 discusses the t distribution.

A key question arises: when calculating the standard deviation and the standard error, which equations should we use? We are trying to estimate the population standard deviation and standard error of the mean based on sample of values obtained through repeat measurements. Consequently, equation 5.7 is preferred for calculating the standard deviation and equation 5.8 for the standard error. Going forward we will adopt, unless stated otherwise, the use of equation 5.7 to estimate the standard deviation of a population and equation 5.8 to estimate the standard error of the mean.

5.4.5   Review of Uncertainty

It can sometimes be difficult to interpret the uncertainty given in the report of an experiment. If, following a study on the mechanical properties of copper, the experimenter states the Young's modulus as $(0.95 \pm 0.05) \times 10^{11}$ Pa, do we take it that the true value for the Young's modulus definitely lies between 0.90×10^{11} Pa and 1.00×10^{11} Pa? How has the uncertainty of 0.05×10^{11} Pa been calculated? If the details of the uncertainty calculation are not given, we are obliged to make an educated guess as to what the uncertainty represents. The interpretation of the uncertainty is straightforward if it is given in terms of the standard error of the mean of the measurements made, and this method is preferred in most situations.

Whatever approach to presenting uncertainties is adopted, sufficient details should be given about the method by which the uncertainty was calculated so that a reader can interpret the uncertainty accordingly.

---

[15] By 'broader distribution' I mean the following. Imagine you were to do an experiment in which you calculated values for $s_{\bar{x}}$ where the sample size you used was small (say $n = 5$). If you did this many times (say 100 times) you could plot a histogram of the distribution of the values of $s_{\bar{x}}$ you had collected. Now repeat this process with much larger sample sizes (say $n = 30$). You would find that the distributions of the standard errors would be broader for $n = 5$ than for $n = 30$. One way to show this is to generate many normally distributed random numbers (e.g. using Excel) and bring them together, first in groups of 5, then in groups of 30. Calculate the $s_{\bar{x}}$'s for the groups of 5 and groups of 30 and plot them separately as histograms.

5.5 Combining Uncertainties When Measurement Errors Are Uncorrelated

In Chapter 4 we used a method for combining uncertainties which, though satisfactory in many situations, tends to overestimate the uncertainty in the calculated quantity.

Suppose that a quantity V depends on quantities a and b, and that the quantity a has error Δa and quantity b has error Δb. It is possible, where there is no correlation between Δa and Δb, for the uncertainty in V to be less that that determined using the methods in Sections 4.5.1 and 4.5.2. We are saying that sometimes Δa can be positive when Δb is negative, and vice versa, but that there is no correlation between the sign and size of Δa and the sign and size of Δb.

Writing V as $V = V(a, b)$, we can show[16] that the variance in the value of V, s_V^2, is related to the variances in a and b, s_a^2 and s_b^2, respectively, in the following way (so long as the errors in a and b are uncorrelated):

$$s_V^2 = \left(\frac{\partial V}{\partial a}\right)^2 s_a^2 + \left(\frac{\partial V}{\partial b}\right)^2 s_b^2.$$

Taking the square root gives

$$s_V = \sqrt{\left(\frac{\partial V}{\partial a}\right)^2 s_a^2 + \left(\frac{\partial V}{\partial b}\right)^2 s_b^2}. \tag{5.9}$$

Equation 5.9 can be rewritten in terms of the standard error in each quantity:

$$s_{\bar{V}} = \sqrt{\left(\frac{\partial V}{\partial a}\right)^2 s_{\bar{a}}^2 + \left(\frac{\partial V}{\partial b}\right)^2 s_{\bar{b}}^2}. \tag{5.10}$$

Recognising that $s_{\bar{V}}$, $s_{\bar{a}}$ and $s_{\bar{b}}$ are taken to be the uncertainties in V, a and b, we can write equation 5.10 alternatively as

$$u_V = \sqrt{\left(\frac{\partial V}{\partial a}\right)^2 u_a^2 + \left(\frac{\partial V}{\partial b}\right)^2 u_b^2}, \tag{5.11}$$

where u_V is the uncertainty in the quantity V, u_a is the uncertainty in the quantity a and u_b is the uncertainty in the quantity b.

---

[16] For more details on combining uncertainties when errors are uncorrelated, see Appendix 2.

Example 2

In an experiment to establish the density of a metal, the mass and volume of a sample were measured several times. Table 5.15 shows the means and standard errors for the mass and volume measurements.

Given that the relationship between density, ρ, mass, m, and volume, V, is $\rho = m/V$, calculate the standard error in ρ using the values in Table 5.15.

Answer

We can write equation 5.9 in terms of ρ, m and V:

$$s_{\bar{\rho}} = \sqrt{\left(\frac{\partial \rho}{\partial m}\right)^2 s_{\bar{m}}^2 + \left(\frac{\partial \rho}{\partial V}\right)^2 s_{\bar{V}}^2}. \tag{5.12}$$

Now,

$$\left(\frac{\partial \rho}{\partial m}\right) = \frac{1}{V}$$

$$= \frac{1}{149 \text{ mm}^3} = 6.711 \times 10^{-3} \text{mm}^{-3}.$$

Note that we evaluate the partial derivative at $V = \bar{V}$:

$$\frac{\partial \rho}{\partial V} = -\frac{m}{V^2}$$

$$= \frac{-1.28 \text{ g}}{(149 \text{ mm}^3)^2} = -5.766 \times 10^{-5} \text{g/mm}^6.$$

Substituting these values into equation 5.12 gives

$$s_{\bar{\rho}} = \sqrt{(6.711 \times 10^{-3} \text{mm}^{-3})^2 \times (0.03 \text{g})^2 + (-5.766 \times 10^{-5} \text{g/mm}^6)^2 \times (6 \text{mm}^3)^2}$$
$$= \sqrt{1.602 \times 10^{-7} \text{g}^2/\text{mm}^6}$$
$$= 4.003 \times 10^{-4} \text{g/mm}^3.$$

The density of the sample is $\rho = m/V = 1.28 \text{ g}/149 \text{ mm}^3 = 8.591 \times 10^{-3} \text{ g/mm}^3$.

The density and the uncertainty in the density can now be quoted to an appropriate number of significant figures as $\rho = (8.6 \pm 0.4) \times 10^{-3} \text{g/mm}^3$.

Table 5.15 Mean mass and volume of a metal sample, including standard errors

Mean mass, $\bar{m}$ (g)	Standard error in $\bar{m}$, $s_{\bar{m}}$ (g)	Mean volume, $\bar{V}$ (mm^3)	Standard error in $\bar{V}$, $s_{\bar{V}}$ (mm^3)
1.28	0.03	149	6

Exercise D

(1) Using the data given in Example 2, calculate the uncertainty in ρ using the method dealing with the addition of uncertainties described in Section 4.5.2.

(2) In an experiment to find the refractive index, n, of a block of glass, the angle of incidence, i, and angle of refraction, r, were measured several times. Table 5.16 shows the means of both quantities and their respective standard errors.

Given that the relationship between n, i and r is $n = \sin i / \sin r$,

  (i) calculate the best estimate for n

  (ii) determine $\partial n / \partial i$ and $\partial n / \partial r$

  (iii) rewrite equation 5.10 in terms of n, i and r

  (iv) use the equation in (iii) to find the standard error in n (note that your calculations require that the angles be expressed in radians, not degrees)

  (v) give the best estimate for n and the standard error in n to the appropriate number of significant figures.

Table 5.16 Means and standard errors of angles of incidence and refraction

Mean angle of incidence, $\bar{i}$ (°)	Standard error in $\bar{i}$, $s_{\bar{i}}$ (°)	Mean angle of refraction, $\bar{r}$ (°)	Standard error in $\bar{r}$, $s_{\bar{r}}$ (°)
61	1	38	2

5.6   Continuous and Discrete Quantities

The quantities we have considered so far vary smoothly or *continuously* and the resolution of our measurements is only limited by the resolution capabilities of the particular instrument being used. Table 5.17 gives examples of physical quantities that vary continuously.

When quantities which vary continuously such as those shown Table 5.17 are measured experimentally, their variation can be described by a continuous distribution, of which the normal distribution is the most widely used example.

Table 5.17 Examples of continuously varying quantities

Continuously varying quantities	time	temperature	length	pressure	mass	force	voltage

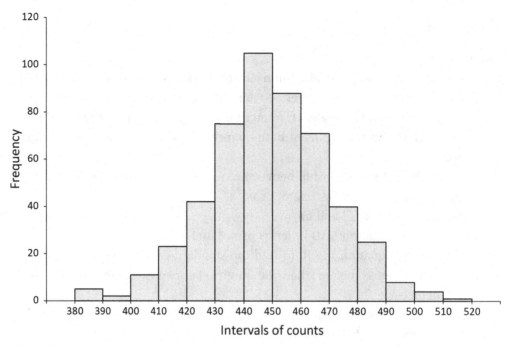

Figure 5.5 Distribution of counts from a radioactivity counting experiment.

However, there is an important class of experiments in which the quantity measured does *not* vary continuously but increases by whole number amounts. We can refer to these as counting experiments. Situations which involve counting include X-ray or radioactivity experiments in which the number of X-rays or emitted particles, respectively, are counted over a period of time.

We can refer to the detection of an X-ray or any other particle as an *event*. If the probability of the event occurring in a time period is small and the occurrence of one event has no influence on later events, then the spread of the number of events occurring in a fixed time period can be described by a discrete probability distribution called the Poisson distribution.[17]

If the number of counts in a fixed time period is recorded in a radioactivity experiment, and this process is repeated many times, a distribution of data emerges which can be presented in the form of a histogram. Figure 5.5 shows a distribution of data for a radioactive counting experiment in which 500 repeat measurements were made.

Note the similarity in the shape of the distribution in Figure 5.5 to that of the normal distribution. In fact, in situations where the number of counts, N, is greater than 10 it is quite acceptable to use the normal distribution as an approximation to the Poisson distribution. Calculating the mean and the standard deviation of the

---

[17] For details of the Poisson distribution see Taylor (1997), chapter 11.

above distribution, we find that mean number of counts $\bar{N} \approx 450$ and the standard deviation of the number of counts $\sigma \approx 21$. This points us to a simple relationship between $\bar{N}$ and σ for the Poisson distribution, which can be justified theoretically. It is that

$$\sigma \approx s = \sqrt{\bar{N}}. \qquad (5.13)$$

If N is the number of counts recorded in a single measurement, then this value must be regarded as our best estimate for the true value of the number of counts, and the standard deviation associated with this value is $\approx \sqrt{N}$.

The standard error of the mean of n repeat measurements, $\sigma_{\bar{N}}$, is written, as before (see equation 5.8),

$$\sigma_{\bar{N}} \approx s_{\bar{N}} = \frac{s}{\sqrt{n}}. \qquad (5.14)$$

Substituting equation 5.13 into equation 5.14 gives

$$\sigma_{\bar{N}} \approx s_{\bar{N}} = \sqrt{\frac{\bar{N}}{n}}. \qquad (5.15)$$

Example 3

In an X-ray experiment the number of X-rays scattered from a powdered sample of nickel oxide over a period of 5 minutes was measured on 10 occasions. Table 5.18 shows the number of counts.

(i)   Find the mean of the counts shown in Table 5.18.
(ii) Calculate the standard deviation, s, of the number of counts.
(iii) Calculate the standard error of the mean value.

Answer

(i) $\bar{N} = 28.0$ counts
(ii) Using equation 5.7, $s = 5.44$ counts
(iii) We can use either equation 5.14 or 5.15 to estimate $\sigma_{\bar{N}}$. Using equation 5.14 we have

$$\sigma_{\bar{N}} \approx s_{\bar{N}} = \frac{s}{\sqrt{n}} = \frac{5.44 \text{ counts}}{\sqrt{10}} = 1.71 \text{ counts}.$$

We can write the number of counts in the five-minute period as

$$N = (28.0 \pm 1.7) \text{ counts}.$$

Table 5.18 Ten values of the number of X-rays emitted from a sample of nickel oxide[a]

19	28	28	36	31	23	31	33	30	21

[a] Each value was obtained by counting the number of X-rays detected over a period of 5 minutes.

5.7   Comment

The statistical methods we have introduced in this chapter for expressing the variability in data due to random errors are powerful and of wide applicability. A significant benefit derived from their use is clarification of what is meant by the uncertainty in a quantity. Limits set by using the standard error of the mean are directly related to the probability of the true value of a quantity lying between those limits.

Whether the method for combining uncertainties discussed in the chapter should be used requires the assumption that errors are uncorrelated is valid. If this assumption is not valid, then the methods for combining uncertainties described between Sections 4.5 and 4.5.3 are preferred. In situations where it is unclear whether the errors are uncorrelated or not, then either method can be used. What is important is that whatever method is chosen it should be clearly described in your laboratory notebook or report.

Problems

5.1. Table 5.19 contains random numbers in the range 0 to 1.
   (i) Take the first two numbers in Table 5.19 and calculate σ and s (given by equations 5.2 and 5.7, respectively). Give σ and s to two significant figures.
   (ii) Repeat (i) but this time calculate σ and s using the first three numbers in Table 5.19, then the first four numbers and so on. At what stage do σ and s agree to one significant figure?

5.2. The oxide $YBa_2Cu_3O_7$ loses oxygen when it is heated above $500\,°C$ in a vacuum. Table 5.20 shows the mass loss from twelve $500\,mg$ samples of the oxide which were heated in a vacuum to a temperature of $800\,°C$ for two hours.
   Using the data in Table 5.20, calculate:
   (i) the sample mean and estimate of the population standard deviation
   (ii) the standard error of the mean
   (iii) the 70% confidence interval for true value of the mass loss.

Table 5.19 Ten random numbers									
0.549	0.022	0.295	0.178	0.190	0.425	0.672	0.996	0.573	0.934

Table 5.20 Mass lost by samples of a ceramic											
Mass loss (mg)	9.5	9.9	8.6	8.8	8.8	9.0	8.6	9.1	9.5	8.3	9.6	9.6

Table 5.21 Blood sugar levels									
Blood sugar level (mmol/L)	4.9	4.8	5.1	5.1	5.0	4.7	5.2	5.3	4.8

5.3. Measurements are made of a person's blood sugar levels. The values obtained are shown in Table 5.21.

Using the data in Table 5.21, calculate:
- **(i)** the sample mean and estimate of the population standard deviation
- **(ii)** the standard error of the mean
- **(iii)** the 95% confidence interval for true value of the blood sugar level.

5.4. A starting pistol is fired. A student standing (352 ± 5) m from the pistol measures the time elapsed between seeing the flash from the pistol and hearing the associated noise. Table 5.22 shows 50 consecutive measurements of the elapsed time.
- **(i)** Present the data in the form of a histogram.
- **(ii)** Do the data look normally distributed?
- **(iii)** Calculate the mean, estimate of population standard deviation and standard error of the mean.
- **(iv)** Given that the velocity of sound is the distance travelled by the sound divided by the elapsed time, calculate the best estimate of the velocity and the uncertainty in the best estimate, assuming errors in distance and time measurements are uncorrelated.

5.5. Table 5.23 shows values obtained through 20 successive measurements of the number of airborne particles in a fixed volume within a clean room.

Using the data in Table 5.23, calculate the 95% confidence interval for the true number of particles in the fixed volume of air.

5.6. The optical density (*o.d.*) of a liquid is given by

$$o.d. = \varepsilon Cl,$$

Table 5.22 Fifty values of elapsed time

Time (s)				
1.24	0.70	1.02	1.07	0.87
1.07	0.87	1.28	1.23	1.10
0.90	1.24	0.82	1.02	1.35
0.96	1.03	1.09	1.31	1.59
1.04	1.07	1.09	1.13	1.36
0.87	1.29	1.34	1.42	0.89
1.53	1.06	1.58	0.98	1.01
1.43	0.80	1.18	1.00	0.74
0.99	0.95	0.97	0.85	1.22
1.10	1.28	1.18	1.16	0.84

Table 5.23 Number of airborne particles in a fixed volume of air

153	132	143	152
159	136	160	165
158	122	149	138
169	170	144	161
147	149	162	150

where ε is called the extinction coefficient, C is the concentration of the absorbing species in the liquid and l is the path length of the light. If the errors in ε, C and l are uncorrelated, and

$$\varepsilon = (15 \pm 1) \, \text{L}/(\text{mol mm})$$
$$C = (0.04 \pm 0.01) \, \text{mol/L}$$
$$l = (1.4 \pm 0.2) \, \text{mm},$$

calculate the best estimate of the optical density and the uncertainty in the best estimate.

5.7. The input voltage, V_{in}, to an amplifier can be calculated using

$$V_{in} = \frac{V_{out} R_f}{R_f + 39 \times 10^3},$$

where V_{out} is the output voltage from the amplifier and R_f is the value of a feedback resistor.

If $R_f = (1.00 \pm 0.05) \times 10^3 \, \Omega$ and $V_{out} = (253 \pm 5) \, \text{mV}$, calculate the best estimate of V_{in} and the uncertainty in the best estimate, assuming that the errors in R_f and V_{out} are uncorrelated.

5.8. The viscosity of a fluid was determined by studying the motion of a sphere falling through the liquid. The viscosity of the fluid, η_f, can be calculated using

$$\eta_f = \frac{\frac{2}{9}r^2 g(\rho_s - \rho_f)}{v_t}.$$

Given that

    r is the radius of the sphere = $(1.2 \pm 0.1)\,\text{mm}$

    g is the acceleration due to gravity = $9.81\,\text{m/s}^2$

    ρ_s is the density of the sphere = $8615\,\text{kg/m}^3$

    ρ_f is the density of the fluid = $1015\,\text{kg/m}^3$

    v_t is the terminal velocity of the sphere through the fluid = $(15 \pm 2)\,\text{mm/s}$,

use this information to calculate the best estimate of the viscosity of the fluid and the uncertainty in the best estimate, assuming errors in the radius and terminal velocity are uncorrelated. Assume the uncertainties in the other quantities are negligible.

5.9. A wooden cantilever is fixed at one end. A mass, M, is attached to the other end of the cantilever. When the mass is displaced and released, it oscillates with a period, T, given by

$$T = 2\pi\sqrt{\frac{4ML^3}{Ebd^3}},$$

where E is the Young's modulus of the wood, L is the length of the cantilever, b is the width of the cantilever and d is its thickness.

  (i) Rearrange the equation to make E the subject of the equation.

      Given that $M = 100.0\,\text{g}$, $T = (0.56 \pm 0.02)\,\text{s}$, $L = (81.4 \pm 0.2)\,\text{cm}$, $b = (1.5 \pm 0.1)\,\text{cm}$ and $d = (0.55 \pm 0.05)\,\text{cm}$.

  (ii) Calculate the best estimate of the Young's modulus of the wood.

 (iii) Assuming there is negligible uncertainty in M, calculate the uncertainty in E assuming the errors in T, L, b and d are uncorrelated.

5.10. An experiment is performed to determine the viscosity of pure water at $20\,^\circ\text{C}$ by allowing water to flow through a narrow hollow tube of length l and internal radius r.

The viscosity of a liquid, η, can be determined using the equation

$$\eta = \frac{kr^4 h}{lQ}.$$

k is a constant = $3.845 \times 10^3\,\text{kg/(m}^2\,\text{s}^2)$ which has negligible uncertainty

$r = (0.39 \pm 0.02)\,\text{mm}$

$l = (10.5 \pm 0.2)\,\text{cm}$

h is a height $= (48.5 \pm 0.1)\,\text{cm}$

Q is the water flow rate of $(0.38 \pm 0.02)\,\text{cm}^3/\text{s}$.

Use these values to calculate the best estimate of the viscosity of water and the uncertainty in the viscosity of water, assuming errors in r, l, h and Q are uncorrelated.

6 Fitting a Line to *x–y* Data Using the Method of Least Squares

6.1 Overview: How Can We Find the Best Line through *x–y* Data?

Linearly related *x–y* data emerge so frequently from science and engineering experiments that the analysis of such data deserves special attention. In general, we seek the quantitative relationship which best describes the dependence of *y* upon *x*. To do this, we need a method by which we can determine the equation of the line that best fits the *x–y* data.

In Chapter 3 we saw that an equation representing the relationship between *x* and *y* quantities[1] can be found by first plotting the data followed by drawing the best straight line through the points on an *x–y* graph (or at least as close as possible to them) with a ruler. The slope, *m*, and intercept, *c*, of this line are calculated and the equation of the line is written $y = mx + c$.

Although positioning a line by eye through *x–y* data is a good way of obtaining reasonable estimates for *m* and *c*, there are several issues with this method:

- No two people draw the same 'best' line through a given data set.
- If the uncertainty in each point on the graph is different, should we take this into account when drawing a line through the points? If the answer is yes, how do we do that?
- Drawing the best line is difficult if the data exhibit large scatter.
- Finding the uncertainties in *m* and *c* directly from the graph (as described in Section 3.3.6) is cumbersome and tends to overestimate their values.

Figure 6.1 shows an example of a graph for which it is difficult to determine the best line through the points.

Both lines in Figure 6.1 appear to fit the data quite well, but which is the better?[2] To answer this we need a tool that avoids the guesswork involved when finding the best line through a set of points by eye. The tool we will use is usually described as

---

[1] Where the *x–y* data look to be (at least approximately) linearly related.
[2] For comparison, the slope and intercept of line 1 are 0.68 and 10, respectively, while for line 2 the slope and intercept are 0.54 and 14, respectively.

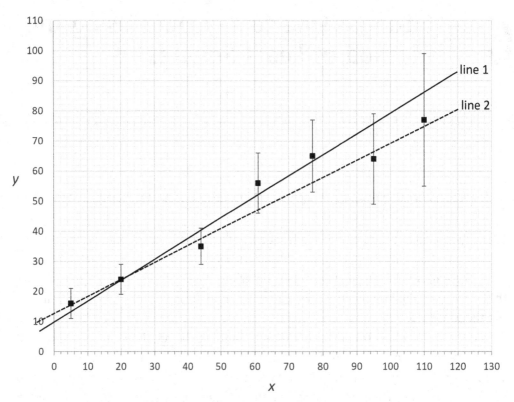

Figure 6.1 Linear *x–y* graph with two 'best' lines through the points.

the fitting of a line to data using the method of *least squares* (also often referred to as *linear regression*).

6.2   The Method of Least Squares

To find the best line through *x–y* data, we begin by assuming that any random errors are confined to measurements of the *y* quantity. This assumption is often valid as it is the *x* quantity that is controlled or adjusted in a stepwise fashion during an experiment, so it is usually possible to know this quantity to high precision. Secondly, we assume that the uncertainty in each measurement of the *y* quantity is the same, which is equivalent to saying that the error bars attached to each point are of the same length. We will deal with a more general situation in which the uncertainties in the *y* values vary from point to point in Section 6.3.

Figure 6.2 shows part of an *x–y* graph with a line passing close to the data points. For a value of *x*, labelled x_i, there are two values of *y* shown on the graph: y_i is the observed value of *y*, i.e. as measured during the experiment; $\hat{y}_i$ refers to the predicted value of *y* found using the equation of a straight line,

$$\hat{y}_i = mx_i + c, \tag{6.1}$$

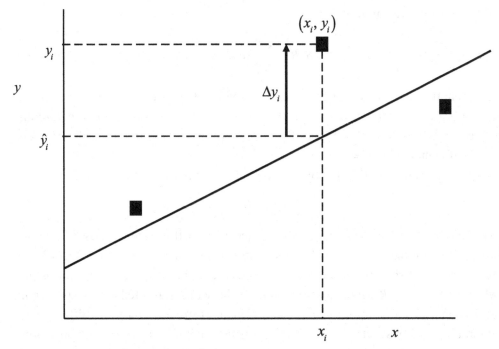

Figure 6.2 x–y graph showing the residual, Δy_i.

where m and c refer to the slope and intercept of the line shown in Figure 6.2.

Δy_i is the difference between the observed and predicted y values and is termed the *residual*, given by

$$\Delta y_i = y_i - \hat{y}_i. \tag{6.2}$$

As we move the line around in an effort to find the position where the line passes closest to the majority of points, Δy_i for each point changes. A criterion is required by which we can decide the best position for the line. This position – and therefore the best values for m and c – is found by applying a theory from statistics called the *principle of maximum likelihood*.[3] This predicts that the best line will be found by minimising the sum of the *squares* of the residuals. Writing the sum of squares of residuals as *SSR*, we can say

$$SSR = (\Delta y_1)^2 + (\Delta y_2)^2 + (\Delta y_3)^2 + \cdots + (\Delta y_n)^2,$$

---

[3] The principle of maximum likelihood considers the probability of obtaining the observed values of x and y during an experiment and asserts that a particular combination of m and c values describes the linear relationship which caused those values to arise. By finding the values of m and c which will make the probability of obtaining the observed set of y values a maximum, the best values of m and c are obtained. For more detail, see Taylor (1997), chapter 8.

which can be abbreviated to

$$SSR = \sum_{i=1}^{i=n} (\Delta y_i)^2. \tag{6.3}$$

The summation indicates that the square of the residuals must be added up for all the data from $i = 1$ to $i = n$, where n is the number of data. To make future equations more compact, we omit the limits of the summation and assume that all sums are calculated from $i = 1$ to $i = n$.

Replacing Δy_i in equation 6.3 by $y_i - \hat{y}_i$ and replacing $\hat{y}_i$ by $mx_i + c$, we can write

$$SSR = \sum (y_i - (mx_i + c))^2. \tag{6.4}$$

We require values for m and c that reduce SSR to the smallest possible value. Those values are taken to be the best estimates of slope and intercept. We could do this by trial and error, but mathematics gives us the tool for finding values for m and c which minimise SSR. First equation 6.4 is partially differentiated with respect to m and the resulting equation set equal to zero. The next step is to partially differentiate equation 6.4 with respect to c and to set the resulting equation equal to zero. Doing this gives

$$\sum x_i(y_i - mx_i - c) = 0 \tag{6.5}$$

and

$$\sum (y_i - mx_i - c) = 0. \tag{6.6}$$

Equations 6.5 and 6.6 are expanded and combined to give the following equations for m and c:

$$m = \frac{n\sum x_i y_i - \sum x_i \sum y_i}{n\sum x_i^2 - (\sum x_i)^2} \tag{6.7}$$

and

$$c = \frac{\sum x_i^2 \sum y_i - \sum x_i \sum x_i y_i}{n\sum x_i^2 - (\sum x_i)^2}. \tag{6.8}$$

Equations 6.7 and 6.8 are extremely useful.[4] In situations where the uncertainties in the y values do not vary from measurement to measurement, and where there are negligible uncertainties in the x values, they allow us to calculate the best values for m and c.

Let us consider an example of the use of equations 6.7 and 6.8.

---

[4] If you have a scientific calculator or a smartphone app that calculates the best estimates for slope and intercept, it is very likely that those calculations are based on equations 6.7 and 6.8.

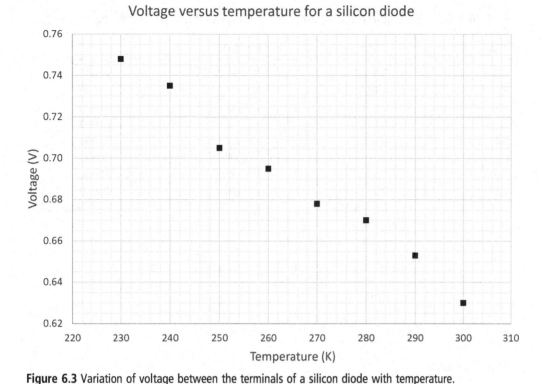

Figure 6.3 Variation of voltage between the terminals of a silicon diode with temperature.

6.2.1 Example of Fitting a Line to *x*–*y* Data

In an experiment to study the behaviour of silicon diodes when they are cooled, the voltage across a diode was measured as a function of the diode temperature. Figure 6.3 shows a graph of the data gathered.

Over the range of temperature shown in Figure 6.3, the relationship between voltage and temperature appears to be linear; therefore we will use equations 6.7 and 6.8 to find the slope and intercept of the best line through the points. We begin by assigning the temperature as the *x* quantity and the voltage as the *y* quantity. The next step is to draw up a table so that the quantities $\sum x_i$, $\sum y_i$, $\sum x_i y_i$ and $\sum x_i^2$ can be calculated.[5] Table 6.1 contains the data (with units included in the column headings) shown plotted in Figure 6.3.

Using equation 6.7 to find the slope of the line,

$$m = \frac{8 \times 1454.42 \text{ K V} - 2120 \text{ K} \times 5.514 \text{ V}}{8 \times 566\,000 \text{ K}^2 - (2120 \text{ K})^2} = \frac{-54.32 \text{ K V}}{33\,600 \text{ K}^2}$$

$$= -1.6167 \times 10^{-3} \text{V/K}.$$

---

[5] In practice, a spreadsheet is usually the most efficient and convenient way to carry out the summations. We will consider the use of a spreadsheet for least squares analysis in Chapter 8.

Table 6.1 Columns required for fitting a line to data using the method of least squares

x_i (K)	y_i (V)	x_iy_i (K V)	x_i^2 (K^2)
300	0.630	189.00	90 000
290	0.653	189.37	84 100
280	0.670	187.60	78 400
270	0.678	183.06	72 900
260	0.695	180.70	67 600
250	0.705	176.25	62 500
240	0.735	176.40	57 600
230	0.748	172.04	52 900
$\Sigma x_i = 2\,120$	$\Sigma y_i = 5.514$	$\Sigma x_iy_i = 1\,454.42$	$\Sigma x_i^2 = 566\,000$

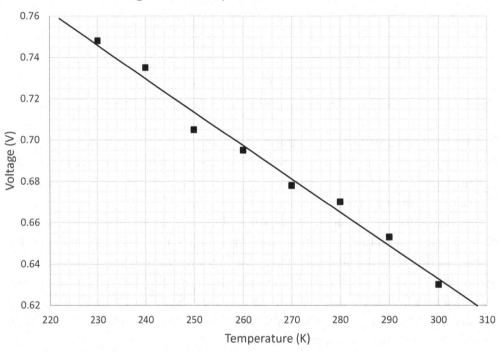

Voltage versus temperature for a silicon diode

Figure 6.4 Best line through the voltage versus temperature data shown in Table 6.1.

Using equation 6.8 to find the intercept of the line,

$$c = \frac{566\,000\ \text{K}^2 \times 5.514\ \text{V} - 2120\ \text{K} \times 1454.42\ \text{K V}}{8 \times 566\,000\ \text{K}^2 - (2120\ \text{K})^2} = \frac{37\,553.6\ \text{K}^2\ \text{V}}{33\,600\ \text{K}^2}$$

$$= 1.1177\ \text{V}.$$

Figure 6.4 shows the best line through the points using the slope and intercept just calculated.

6.2.2   Possible Calculation Difficulties

When finding m and c using equations 6.7 and 6.8, care must be taken not to round the results of intermediate calculations, as this can greatly influence the values of m and c. For example, the numerator in the calculation for m in Section 6.2.1 is $-54.32\,KV$. If each value in the numerator is rounded to three significant figures (for example $1454.42\,KV$ is rounded to $1450\,KV$), then the numerator becomes $-81.2\,KV$. Rounding has a big effect when terms which are about the same size are subtracted in the numerator (or the denominator), as in this example. It is good advice to keep any intermediate results to as many figures as possible and only round values at the end. The problem is minimised if fitting is performed using a scientific calculator, spreadsheet or a calculator app for a smartphone. Each of these hold numbers internally to many figures, so the influence of rounding on the final values of m and c is minimal.

Exercise A

(1) The slope of a straight line was calculated as follows:

$$m = \frac{10 \times 672.93 - 2350.6 \times 2.785}{10 \times 552\,900 - (2350.6)^2}.$$

  (i) Calculate m using the numbers as shown. Give m to two significant figures.

  (ii) Round each number in the calculation to three significant figures and recalculate m. Give m to two significant figures.

(2) Table 6.2 shows a set of x–y data. Assuming that y is linearly related to x, find the slope and intercept of the best line through the data using the method of least squares. Give the slope and intercept to four significant figures.

Table 6.2 Linearly related x–y data

x	y
2.1	45.4
4.4	65.7
6.3	73.4
8.3	95.0
10.2	102.8
12.3	121.2
14.6	134.7
16.7	155.3

Table 6.3 Variation of acceleration due to gravity with height above the Earth's surface

h (m)	g' (m/s^2)
1×10^4	9.76
2×10^4	9.74
3×10^4	9.70
4×10^4	9.69
5×10^4	9.66
6×10^4	9.62
7×10^4	9.59
8×10^4	9.55
9×10^4	9.54
10×10^4	9.51

(3) The acceleration due to gravity, g', was measured at various heights, h, above the Earth's surface. Table 6.3 contains the values of g' corresponding to various heights.

   (i) Taking height as the x variable and the acceleration due to gravity as the y variable, calculate the slope and intercept using the least-squares method. Take the relationship between g' and h as $g' = g(1 - 2h/R_E)$,[6] where R_E is the radius of the Earth and g is the acceleration due to gravity on the surface of the Earth.

   (ii) Rearrange the equation into the form $y = mx + c$. Use your values for m and c to find values for R_E and g.

6.2.3   Standard Errors in the Estimates of the Slope and Intercept

We can find the best values for the slope and intercept of a line through a set of $x–y$ data using equations 6.7 and 6.8. However, it is not possible to decide how many figures m and c should be quoted to until we have established the uncertainties in m and c.

   We take these uncertainties to be the standard errors in m and c, which we write as s_m and s_c, respectively. To calculate s_m and s_c we recognise the following:

   (i)   For each value of x, the corresponding value of y has some error.

  (ii) The error in each value of y contributes to the standard errors in m and c.

 (iii) If the errors in y are uncorrelated, we can adapt equation 5.10 to give the standard error in m as

---

[6] This equation is valid when h is much less than R_E.

$$s_m = \sqrt{\left(\frac{\partial m}{\partial y_1}\right)^2 s_1^2 + \left(\frac{\partial m}{\partial y_2}\right)^2 s_2^2 + \cdots + \left(\frac{\partial m}{\partial y_n}\right)^2 s_n^2}, \tag{6.9}$$

where s_1, s_2, ..., s_n are the standard deviations in the observed values y_1, y_2, ..., y_n, respectively. If the standard deviation in the y values is constant, then s_1, s_2, ..., s_n can each be replaced by s, allowing equation 6.9 to be written

$$s_m = \left[s^2 \sum \left(\frac{\partial m}{\partial y_i}\right)^2\right]^{1/2}. \tag{6.10}$$

A similar equation to equation 6.10 can be written for the standard error in c by replacing m in equation 6.10 by c:

$$s_c = \left[s^2 \sum \left(\frac{\partial c}{\partial y_i}\right)^2\right]^{1/2}. \tag{6.11}$$

Several steps are required before we can arrive at explicit equations for s_m and s_c. We will not go through the steps here, but simply quote the results:

$$s_m = \frac{s n^{1/2}}{\left[n \sum x_i^2 - \left(\sum x_i\right)^2\right]^{1/2}} \tag{6.12}$$

$$s_c = \frac{s \left(\sum x_i\right)^{1/2}}{\left[n \sum x_i^2 - \left(\sum x_i\right)^2\right]^{1/2}}, \tag{6.13}$$

where s is the standard deviation in each y value of the data point:

$$s = \left[\frac{1}{n-2} \sum (y_i - m x_i - c)^2\right]^{1/2}. \tag{6.14}$$

The equation for s is similar to equation 5.7 for estimating the standard deviation of a population based on a sample of data drawn from that population.[7] The deviation of each point with respect to the line is $y_i - m x_i - c$. The deviation is squared and summed for all data points. The reason that $n-2$ appears in the denominator of equation 6.14 is discussed in Appendix 1.

6.2.4 Example of a Calculation of the Uncertainty in m and c

We will use equations 6.12 to 6.14 to calculate the uncertainty in m and the uncertainty in c for the data given in Section 6.2.1. Fitting by least squares gave m and c as

---

[7] See Bevington and Robinson (2003), chapter 6.

Table 6.4 Columns required for calculation of uncertainty in m and c

x_i (K)	y_i (V)	$y_i - mx_i - c$ (V)	$(y_i - mx_i - c)^2$ (V^2)
300	0.630	-2.6667×10^{-3}	7.1111×10^{-6}
290	0.653	4.1667×10^{-3}	1.7361×10^{-5}
280	0.670	5.0000×10^{-3}	2.5000×10^{-5}
270	0.678	-3.1667×10^{-3}	1.0028×10^{-5}
260	0.695	-2.3333×10^{-3}	5.4444×10^{-6}
250	0.705	-8.5000×10^{-3}	7.2250×10^{-5}
240	0.735	5.3333×10^{-3}	2.8444×10^{-5}
230	0.748	2.1667×10^{-3}	4.6944×10^{-6}

$$\Sigma \, (y_i - mx_i - c)^2 = 1.7033 \times 10^{-4}$$

$$m = -1.6167 \times 10^{-3} \text{ V/K}$$
$$c = 1.1177 \text{ V}.$$

Table 6.4 has been drawn up so that s can be calculated.

Using equation 6.14,

$$s = \left(\frac{1.7033 \times 10^{-4} \text{ V}^2}{8 - 2} \right)^{\frac{1}{2}} = 5.3281 \times 10^{-3} \text{ V}.$$

s_m and s_c are calculated using equations 6.12 and 6.13. $n\sum x_i^2 - \left(\sum x_i\right)^2$ was calculated for this data set in Section 6.2.1 and found to be equal to $33\,600\,\text{K}^2$. It follows that

$$s_m = \frac{5.3281 \times 10^{-3} \text{ V} \times 8^{\frac{1}{2}}}{\left(33\,600 \text{ K}^2\right)^{\frac{1}{2}}} = 8.2214 \times 10^{-5} \text{ V/K}.$$

To calculate s_c, we need $\sum x_i^2$, which is given as $566\,000\,\text{K}^2$ in Table 6.1. It follows that

$$s_c = \frac{5.3281 \times 10^{-3} \text{ V} \times \left(566\,000 \text{ K}^2\right)^{\frac{1}{2}}}{\left(33\,600 \text{ K}^2\right)^{\frac{1}{2}}} = 0.021868 \text{ V}.$$

We are now able to quote m and c to the number of figures consistent with the uncertainties in each of these quantities:

$$m = (-1.62 \pm 0.08) \times 10^{-3} \text{ V/K}$$
$$c = (1.12 \pm 0.02) \text{ V}.$$

As usual, the uncertainties have been rounded to one significant figure.

Exercise B

Refer to questions (2) and (3) of Exercise A in this chapter. Calculate the uncertainty in the slope and intercept for the data given in each question. Quote the slope and intercept, along with their uncertainties to an appropriate number of significant figures.

6.2.5   Interpretation of Uncertainties in m and c

s_m and s_c, which we take to be the uncertainties in m and c, respectively, are the standard errors of these quantities. We saw in Section 5.4.3 that the true (or population) mean of a distribution of numbers has about a 70% chance of lying within one standard error of the sample mean and about a 95% chance of lying within two standard errors of the sample mean. Similarly, when quoting s_m, we are saying that the true value of the slope has about a 70% chance of lying within the interval $m - s_m$ to $m + s_m$ and about a 95% chance of lying within the interval $m - 2s_m$ to $m + 2s_m$. Similarly, the true value for the intercept has about a 70% chance of lying within the interval $c - s_c$ to $c + s_c$.

6.3   Weighting the Fit

To account for situations in which the uncertainties in the y values vary from point to point, we use *weighted* least squares when fitting a line to data.[8] The sum of squares is weighted such that, when fitting takes place, the calculated line lies closest to those points with least uncertainty, in effect favouring those points. Situations in which a weighted fit is required are quite common and include those where:

- changes in resolution occur while making measurements with an instrument which is switched between operating ranges during the course of an experiment. For example, a voltmeter might be used during an experiment in an experiment on its 200 mV range with a corresponding resolution uncertainty of 0.1 mV. If sometime later the voltmeter is switched to the 2 V range, the resolution uncertainty becomes 1 mV.
- y values must be transformed so that a straight-line relationship can be produced.

Figure 6.5 shows linearly related data in which the uncertainties in the y values are not constant. Whatever the reason for the variation in the size of the

---

[8] Note that weighted fitting is generally not available on scientific calculators nor smartphone apps.

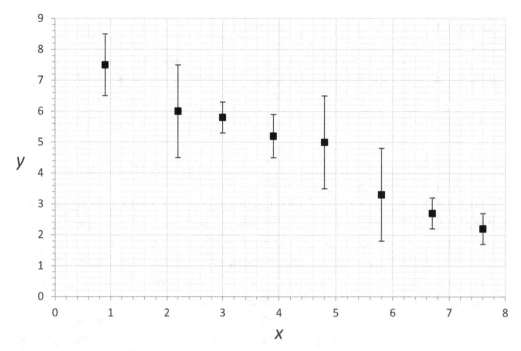

Figure 6.5 Linearly related *x–y* data in which the uncertainties in the *y* values vary.

uncertainties, we must make the most of those points that have the smallest uncertainty, and give less weight to those with large uncertainty.

When uncertainties vary from point to point, the standard deviation in the *i*th value (written as s_i) must be used in the calculations of m, s_m, c and s_c.

In order to be able to write the equations for m, c etc. in a more condensed form, we introduce the quantity Δ, given by

$$\Delta = \sum \frac{1}{s_i^2} \sum \frac{x_i^2}{s_i^2} - \left(\sum \frac{x_i}{s_i^2} \right)^2. \tag{6.15}$$

The remaining quantities can be written as[9]

$$m = \frac{\sum \frac{1}{s_i^2} \sum \frac{x_i y_i}{s_i^2} - \sum \frac{x_i}{s_i^2} \sum \frac{y_i}{s_i^2}}{\Delta} \tag{6.16}$$

$$s_m = \left(\frac{\sum \frac{1}{s_i^2}}{\Delta} \right)^{½} \tag{6.17}$$

---

[9]   See Bevington and Robinson (2003), chapter 6, for derivations of equations 6.15 to 6.19.

$$c = \frac{\sum \frac{x_i^2}{s_i^2} \sum \frac{y_i}{s_i^2} - \sum \frac{x_i}{s_i^2} \sum \frac{x_i y_i}{s_i^2}}{\Delta} \tag{6.18}$$

$$s_c = \left(\frac{\sum \frac{x_i^2}{s_i^2}}{\Delta} \right)^{1/2}. \tag{6.19}$$

Applying equations 6.15 through 6.19 to x–y data requires a fair amount of work. In this situation a spreadsheet is of great assistance, as discussed in Chapter 8.

6.3.1 Example of a Weighted Fit

In an experiment to study the deposition of copper, the mass of a copper electrode was measured while an electrical current was passed through a solution of copper sulfate surrounding the electrode. The relationship between mass, M, and time, t, is

$$M = kt + M_0,$$

where M_0 is the mass of the copper electrode at $t = 0$ and k is a constant.

When the electrode mass was below 50 g a balance was used capable of resolving down to ± 0.1 g. When the mass exceeded 50 g, a balance with a larger range was used with a resolution of ± 0.5 g. Table 6.5 shows data gathered of mass, M, as a function of time, t, and the corresponding graph is shown in Figure 6.6. As the error bars are not constant, a weighted fit is required.

To assist in applying equations 6.15 to 6.19, Table 6.6 has been drawn up, containing all the relevant quantities.

We can now calculate Δ using equation 6.15:

$$\Delta = 316\,\text{g}^{-2} \times 2\,032\,800\ \text{min}^2/\text{g}^2 - \left(19\,600\ \text{min}/\text{g}^2 \right)^2$$
$$= 258\,204\,800\ \text{min}^2/\text{g}^4.$$

Table 6.5 Variation of the mass of a copper electrode with time in an electrolysis experiment

t (min)	M (g)
10	47.7 ± 0.1
60	48.7 ± 0.1
90	49.4 ± 0.1
160	50.4 ± 0.5
200	51.6 ± 0.5
240	52.1 ± 0.5
300	52.7 ± 0.5

Table 6.6 Weighted fit of data in Table 6.5

x_i (min)	y_i (g)	s_i (g)	$1/s_i^2(\mathrm{g}^{-2})$	$x_i/s_i^2(\mathrm{min/g}^2)$	$y_i/s_i^2(\mathrm{g}^{-1})$	$x_iy_i/s_i^2(\mathrm{min/g})$	$x_i^2/s_i^2(\mathrm{min}^2/\mathrm{g}^2)$
10	47.7	0.1	100	1 000	4 770	47 700	10 000
60	48.7	0.1	100	6 000	4 870	292 200	360 000
90	49.4	0.1	100	9 000	4 940	444 600	810 000
160	50.4	0.5	4	640	201.6	32 256	102 400
200	51.6	0.5	4	800	206.4	41 280	160 000
240	52.1	0.5	4	960	208.4	50 016	230 400
300	52.7	0.5	4	1 200	210.8	63 240	360 000
			$\sum\dfrac{1}{s_i^2}=316$	$\sum\dfrac{x_i}{s_i^2}=19\,600$	$\sum\dfrac{y_i}{s_i^2}=15\,407.2$	$\sum\dfrac{x_iy_i}{s_i^2}=971\,292$	$\sum\dfrac{x_i^2}{s_i^2}=2\,032\,800$

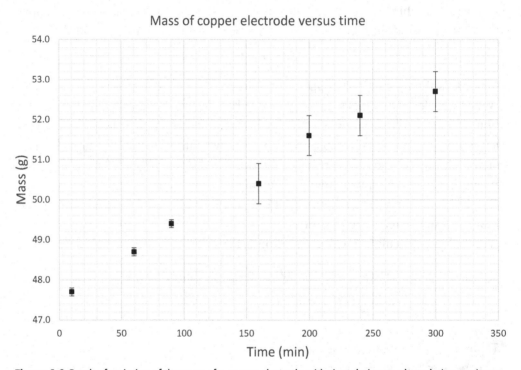

Figure 6.6 Graph of variation of the mass of a copper electrode with time during an electrolysis experiment.

m can be found using equation 6.16:

$$m = \frac{316\ \mathrm{g}^{-2} \times 971\,292\ \mathrm{min/g} - 19\,600\ \mathrm{min/g}^2 \times 15\,407.2\ \mathrm{g}^{-1}}{258\,204\,800\ \mathrm{min}^2/\mathrm{g}^4}$$

$$= \frac{4\,947\,152\ \mathrm{min/g}^3}{258\,204\,800\ \mathrm{min}^2/\mathrm{g}^4}$$

$$= 1.9160 \times 10^{-2}\mathrm{g/min}.$$

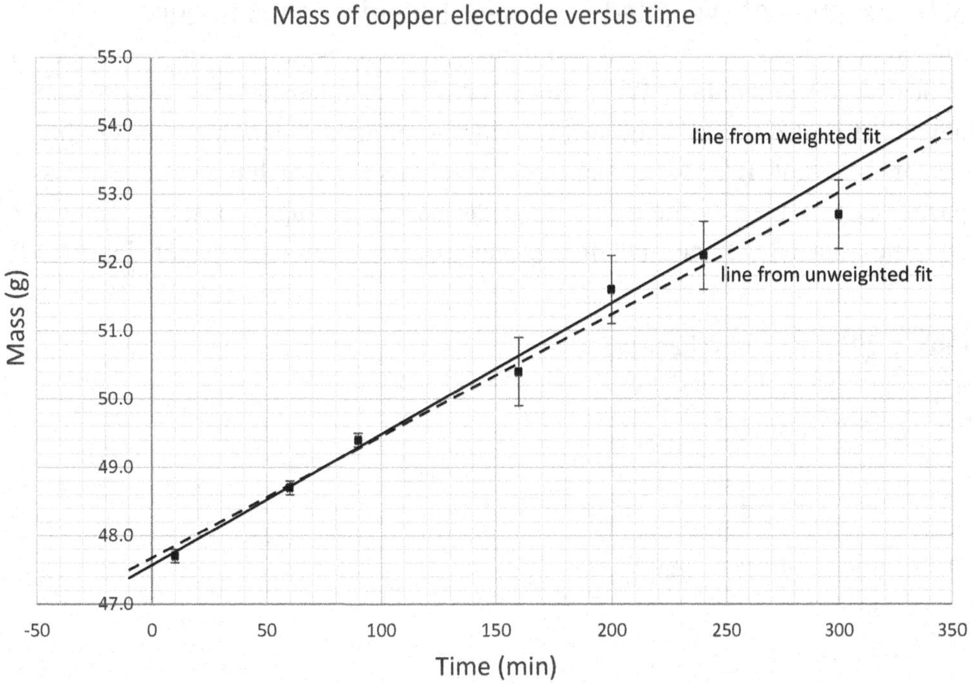

Figure 6.7 Graph showing best fit lines using weighted and unweighted least squares.

s_m can be found using equation 6.17:

$$s_m = \left(\frac{316 \text{ g}^{-2}}{258\ 204\ 800 \quad \text{min}^2/\text{g}^4} \right)^{\frac{1}{2}} = 1.11 \times 10^{-3} \text{g/min}.$$

The slope is written as $m = (1.92 \pm 0.11) \times 10^{-2}$ g/min.

Similarly, c and s_c can be found using equations 6.18 and 6.19, respectively:

$$c = \frac{2\ 032\ 800 \quad \text{min}^2/\text{g}^2 \times 15\ 407.2 \text{ g}^{-1} - 19\ 600 \text{ min/g}^2 \times 97\ 1292 \quad \text{min/g}}{258\ 204\ 800 \quad \text{min}^2/\text{g}^4}$$

$$= 47.5686 \text{ g}$$

$$s_c = \left(\frac{2\ 032\ 800 \quad \text{min}^2/\text{g}^2}{258\ 204\ 800 \quad \text{min}^2/\text{g}^4} \right)^{\frac{1}{2}} = 8.87 \times 10^{-2} \text{ g},$$

so that c can be written as $c = (47.57 \pm 0.09)$ g.

A line of best fit can now be added to the data, as shown in Figure 6.7. For comparison, this graph shows the line of best fit using an unweighted fit.

Exercise C

Perform an *unweighted* fit to the data in Table 6.5 and show that the slope and intercept using this method are $m = (1.78 \pm 0.10) \times 10^{-2}$ g/min and $c = (47.7 \pm 0.18)$ g.

6.3.2   Example of Weighted Fit Where Linearisation is Required

In the previous example, the equation relating the data is of the form $y = mx + c$, so no linearisation is required. When linearisation *is* required it often means that a weighted fit is also needed – even if the raw experimental data have uncertainties which are constant. Take, for example, an experiment in which the voltage across a capacitor is measured as the capacitor is discharged through a resistor. Table 6.7 shows the data of voltage and time gathered during the experiment. Figure 6.8

Table 6.7 Variation of voltage across a capacitor with time as the capacitor discharges	
Time (s)	Voltage (V)
5	10.9 ± 0.1
10	6.5 ± 0.1
15	4.1 ± 0.1
20	2.3 ± 0.1
25	1.5 ± 0.1
30	0.8 ± 0.1
35	0.5 ± 0.1
40	0.4 ± 0.1
45	0.2 ± 0.1
50	0.2 ± 0.1

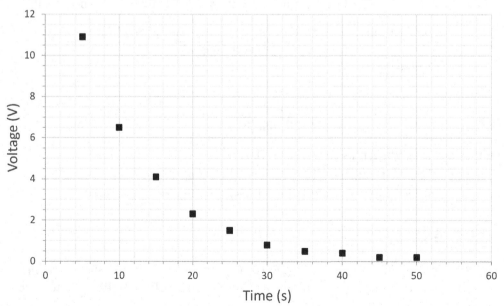

Figure 6.8 Variation of voltage across a discharging capacitor with time.

shows a plot of the data (as the error bars are too small to show clearly, they have been omitted from the graph).

The relationship between the voltage across a capacitor, V, and the time, t, as the capacitor discharges is

$$V = V_0 \exp\left(-\frac{t}{\tau}\right), \tag{6.20}$$

where V_0 is the voltage at time $t = 0$ and τ is the time constant for the discharge. Note that in Table 6.7 the uncertainty in the voltage measurements is constant at 0.1 V. However, equation 6.20 must be transformed before a straight line can be fitted to the data, and the uncertainty in the transformed quantity is *not* constant.

To transform equation 6.20 into an equation of the form $y = mx + c$, we take the natural logarithms of both sides of the equation. This gives

$$\ln(V) = \ln(V_0) - \frac{t}{\tau}. \tag{6.21}$$

Plotting $\ln(V)$ against t should give a straight line with a slope of $-1/\tau$ and an intercept of $\ln(V_0)$. The quantity plotted on the y axis is $\ln(V)$. If the uncertainty in V is 0.1 V, what is the uncertainty in $\ln(V)$? To answer this, we use the mathematical tool of partial differentiation, which we applied in Chapter 4. If $y = f(V)$, then $\Delta y = (\partial y / \partial V)\Delta V$. In this example, $y = \ln(V)$, so $\partial y / \partial V = 1/V$. It follows that $\Delta y = \Delta V / V$.

In this chapter we have used the symbol s_i to represent the standard deviation in the ith y value, so that in this example $s_i = \Delta V_i / V$. Now we can draw up a table similar to Table 6.6 so that the slope and intercept of the $\ln(V)$ versus t data using weighted least squares can be found.

Exercise D

(1) To fit equation 6.21 to the data in Table 6.7 requires a table similar to Table 6.6 to be created. Table 6.8 shows the beginning of such a table.
  (i) Complete Table 6.8 using the data in Table 6.7.
  (ii) Plot a graph of $\ln(V)$ versus t using the data in Table 6.8.

Table 6.8 Columns required for the calculation of slope and intercept (and uncertainties in these quantities) for the data given in Table 6.7

x_i(s)	$y_i = \ln(V_i)$	$s_i = \Delta V_i / V_i$	$1/s_i^2$	x_i/s_i^2(s)	y_i/s_i^2	$x_i y_i/s_i^2$(s)	x_i^2/s_i^2(s^2)
5	2.388763	0.009174	11 881	59 405	28 380.89	141 904.5	297 025
10	1.871802	0.015385	4 225	42 250	7 908.364	79 083.64	422 500
15	1.410987	0.02439	1 681	25 215	2 371.869	35 578.04	378 225

Table 6.9 Variation of counts per second with time for radiation emitted from a radioactive source	
t (s)	I (counts/s)
60	1 580
120	860
180	460
240	290
300	130
360	67
420	53
480	16

   (iii)   Calculate the slope and intercept of the line.

   (iv)   Calculate the uncertainties in the slope and intercept.

   (v) Find V_0 and τ and the associated uncertainties in these quantities.

(2) The data in Table 6.9 were gathered in an experiment to study the rate of decay of a radioactive element.

     The relationship between number of counts per second, I, and the time, t, is $I = I_0 \exp(-t)$, where I_0 is the number of counts per second at time $t = 0$ and λ is the decay constant of the radioactive source.

  (i)   Linearise the equation and plot a graph of the linearised data.

  (ii) Taking the uncertainty in I to be equal to $\sqrt{I}$, carry out a weighted fit to data to find I_0 and λ and the uncertainties in these quantities.

  (iii) Show the line of best fit on the graph.

(3) In an optics experiment, the diameter, D, of several interference rings was measured. Table 6.10 shows the results obtained.

     The relationship between diameter, D, and ring number, n, is $D = \sqrt{An + B}$, where A and B are constants.

  (i)   Linearise the above equation.

  (ii) Perform an unweighted least-squares fit to the data in Table 6.10 to find A and B and the uncertainties in A and B.

  (iii) Perform a weighted least-squares fit to the data in Table 6.10 to find A and B and the uncertainties in A and B.

  (iv) Calculate the percentage difference between the values for A obtained using weighted and unweighted least squares.

  (v) Calculate the percentage difference between the values for B obtained using weighted and unweighted least squares.

Table 6.10 Diameter of interference rings as a function of ring number	
n	D (mm)
10	2.30 ± 0.02
11	2.35 ± 0.02
12	2.42 ± 0.02
13	2.50 ± 0.02
14	2.55 ± 0.02
15	2.61 ± 0.02
16	2.64 ± 0.02
17	2.71 ± 0.02
18	2.75 ± 0.02

6.4 How Well Does the Line Fit the x–y Data? The Linear Correlation Coefficient, r

An examination of x–y data, presented as a graph, is the most direct way to establish whether the data lie along a straight line, whether the data are subject to large scatter and whether one or more outliers exist.

If the fit is good, the data should be scattered randomly about the line of best fit. If there is an obvious trend in the scatter of the data – for example if the data consistently lie below the line of best fit at small values of x and above the line at large values of x – then some consideration should be given to whether a linear relationship really exists between the quantities plotted on the x and y axes.

There are several ways in which the 'goodness of fit' of the line to the data can be established. A useful statistic which quantifies the extent to which y is linearly related to x is the linear correlation coefficient, r. This statistic is found on most computer packages and calculators capable of fitting a straight line by least squares. r is given by

$$r = \frac{n\sum x_i y_i - \sum x_i \sum y_i}{\left[n\sum x_i^2 - \left(\sum x_i\right)^2\right]^{1/2}\left[n\sum y_i^2 - \left(\sum y_i\right)^2\right]^{1/2}}, \qquad (6.22)$$

and r ranges from 1 to -1.

An r of 1 indicates perfect correlation between x and y values, with y increasing as x increases (i.e. the slope of the line has a positive sign). An r of -1 also indicates perfect correlation between x and y values, but in this case y *decreases* as x increases (i.e. the slope of the line has a negative sign). The closer r is to zero, the less convincing is the correlation between x and y.

Example 1

Table 6.11 contains two sets of y data which are linearly related to the x values. One set of data has (unrealistically) no scatter. The other has scatter typical of data gathered in an experiment in which the y values are subject to random errors. These values and the lines of best fit are displayed in Figure 6.9.

Table 6.11 Two sets of *x*–*y* data

x	y (no scatter)	y (with random scatter)
2	25.1	32.1
4	28.2	39.1
6	31.3	40.4
8	34.4	46.4
10	37.5	43.7
12	40.6	54.0

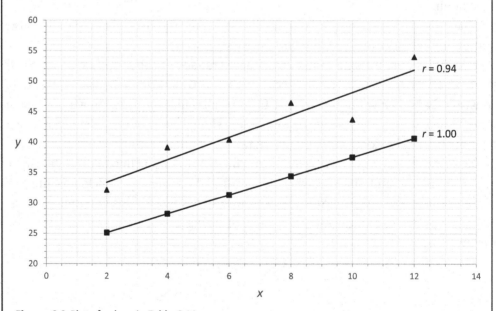

Figure 6.9 Plot of values in Table 6.11.

As an indicator of goodness of fit, r should be used cautiously, as a 'good' value for the magnitude of r, say $|r| \approx 0.8$, can be obtained when there are only a few values (say six or fewer pairs of x–y values), when there is in fact no real correlation between the x and y quantities.[10]

Exercise E

Determine the correlation coefficient for the diode data in Table 6.1.

6.5   Comment

This chapter has introduced the technique of fitting lines to linearly related x–y data by the method of least squares. The technique can be extended to situations where equations have more than two parameters, such as

$$y = A \ln (x) + Bx + C, \text{ or } y = A + Bx^c,$$

where A, B and C are the parameters.

In this chapter, discussion has been confined to cases where there is a linear relationship between x and y and the errors in measured quantities are limited to the y quantity. The least-squares functions found on most calculators and in computer packages adopt the same assumption regarding the errors. The simple fact is that when the errors in the x quantities are *not* negligible, deriving equations for the slope and intercept becomes more challenging. The difficulty of fitting lines to data when there are errors in both the x and y quantities has been recognised for many years. Macdonald and Thompson (1992) review of a variety of approaches for finding slope and intercept when both x and y values are subject to error.

To end this chapter on fitting a line to data using the method of least squares, it is worth making two remarks:

(i) Fitting a line to data using least squares is a powerful data analysis technique, but effort is required to ensure that the data substituted into the equations are worth analysing in the first place. Data with large scatter are always difficult to deal with, no matter how sophisticated the analysis tool, and it is possible that in some cases the nature of the underlying relationship cannot be established

---

[10] A discussion of what constitutes a good fit of a line to data can be found in Kirkup (2012), chapter 6.

clearly. Therefore, every effort is required to minimise errors at the data gathering stage and to avoid relying on the analysis technique to 'save the day'.

(ii) The importance of inspecting x–y data by graphing the data cannot be overstated. It is easy to enter data incorrectly into a calculator or computer package which ultimately produces ridiculous values for the slope and intercept. These can go unrecognised unless they are compared with approximate values obtained by plotting a graph and estimating the slope and intercept of a line drawn through the points with a ruler. Computers are essential when fast or repetitive calculations are required, but cannot match the eye/brain combination when it comes to spotting patterns, trends and anomalies.

Problems

6.1. Calculate:

    (i) the slope and intercept of the x–y data shown in Table 6.12 using unweighted least squares

   (ii) the uncertainty in the slope and intercept

  (iii) the linear correlation coefficient, r.

6.2. In an experiment, the optical behaviour of a material was studied. The intensity of light, I, reflected from a crystal of lithium fluoride was measured as the angular position, θ, of a polariser placed between the light source and the crystal was changed. Table 6.13 gives data of the intensity (in arbitrary units) as a function of angle.

Assume that the function that describes the data in Table 6.13 is of the form $I = (I_{max} - I_{min}) \cos(2\theta) + I_{min}$, where I_{max} and I_{min} are constants.

    (i) What would you plot to obtain a straight line for the data shown in Table 6.13?

Table 6.12 Linearly related *x*–*y* data

x	y
45	22
39	36
31	38
24	48
18	55
11	64
4	76

Table 6.13 Variation of intensity of light reflected from a lithium fluoride crystal as a function of the angular position of a polariser

θ (degrees)	Intensity (arbitrary units)
0	1.86
20	1.63
40	1.13
60	0.52
80	0.16
100	0.00
120	0.57
140	1.11
160	1.56
180	1.71

Table 6.14 Mass of oxide formed as a function of time in a corrosion experiment

Time, t (h)	Mass, M (mg)
0.5	6 ± 1
1	9 ± 1
2	12 ± 1
3	15 ± 1
4	17 ± 1
5	18 ± 1
6	20 ± 1
7	20 ± 1
8	22 ± 1

(ii) Fit a straight line to the data and determine the slope, m, and intercept, c, assuming an unweighted least-squares fit is appropriate. Give m and c to three significant figures.

(iii) Calculate the uncertainty in m and c. Give m and c and associated uncertainties to an appropriate number of significant figures.

(iv) Using the slope and intercept, calculate I_{max} and I_{min}.

6.3. In a corrosion experiment, the mass of oxide formed on a metal is measured as a function of time during which the metal is exposed to air. Table 6.14 gives the data gathered during the experiment.

Assume that the relationship between mass, M, and time, t, can be written $M = \sqrt{kt + D}$, where k and D are constants.

Table 6.15 Variation of mass of gas adsorbed per unit area as a function of pressure	
g (kg/m^2)	p (N/m^2)
1.40×10^{-4}	0.28
1.76×10^{-4}	0.40
2.21×10^{-4}	0.61
2.78×10^{-4}	0.95
3.28×10^{-4}	1.70
3.84×10^{-4}	3.40

(i) Linearise the equation, indicate what you would plot in order to obtain a straight line and state clearly how the slope and intercept of the line are related to k and D.

(ii) Explain why a weighted fit of a straight line to the data in Table 6.14 would be preferred to an unweighted fit.

(iii) Using the appropriate weighting, find the slope and intercept of the best line through the transformed data.

6.4. The relationship between the mass of gas adsorbed per unit area of surface, g, and the gas pressure, p, can be written (assuming constant temperature)

$$\frac{p}{g} = \frac{1}{n} + \frac{s}{n}p,$$

where s and n are constants. Table 6.15 gives experimental data obtained from an adsorption experiment carried out at $0\,^\circ$C.

Use unweighted least squares to determine the values of n and s. Hint: first rearrange the equation into the form $y = mx + c$.

6.5. Measurements are made of the volume, V, occupied by a liquid at temperature θ. The values recorded are shown in Table 6.16.

Assume that $V = 1 + B\theta + D\theta^2$, where B and D are constants.

(i) Linearise the equation $V = 1 + B\theta + D\theta^2$.

(ii) Use the method of least squares to find values for B and D and the uncertainties in these quantities, assuming that the uncertainties in volume values are constant and that there is negligible uncertainty in the temperature measurement.

6.6. The relationship between the current, I, and the voltage, V, of a semiconductor diode can be written $I = I_0 \exp\left({eV}/{nkT}\right)$, where

e is the magnitude of the charge on an electron ($= 1.60 \times 10^{-19}$ C)

k is Boltzmann's constant ($= 1.38 \times 10^{-23}$ J/K)

Table 6.16 Volume occupied by a liquid as the temperature of the liquid increases

V (cm^3)	θ (°C)
1.032	10.0
1.063	20.0
1.094	29.5
1.125	39.5
1.156	50.0
1.186	60.5
1.215	69.5
1.244	79.5
1.273	90.0
1.300	99.0

Table 6.17 Experimental current and voltage values for a semiconductor diode

I (A)	V (volts)
7.53×10^{-5}	0.50
3.17×10^{-4}	0.55
1.07×10^{-3}	0.60
3.75×10^{-3}	0.65
1.35×10^{-2}	0.70
4.45×10^{-2}	0.75
1.75×10^{-1}	0.80
5.86×10^{-1}	0.85

T is the temperature in kelvin

I_0 is the saturation current (which is a constant)

n is a constant called the *ideality factor*.

Table 6.17 shows a set of data of current as a function of voltage when the diode is maintained at a temperature of 300 K.

(i) Linearise $I = I_0 \exp\left(^{eV}/_{nkT}\right)$.

The uncertainty in each value of current in Table 6.17 is equal to 2% of that current.

(ii) Should a weighted or unweighted least-squares fit to data be performed? Justify your answer.

Table 6.18 Time delay of detection of sound wave as a function of microphone position

d (m)	Time delay, *t* (ms) ± 0.05 ms
15.0	35.36
20.0	36.30
25.0	37.08
30.0	38.22
35.0	39.69
40.0	41.32
45.0	42.90
50.0	44.50

Table 6.19 Variation of resistance of a graphite rod with distance between electrodes

l (mm)	*R* (mΩ)
38	3.92
67	6.27
94	8.53
119	10.59
137	13.14
162	14.17
185	16.16
227	19.68

(iii) Perform the appropriate fit to find I_0 and n.

(iv) Calculate the uncertainties in I_0 and n.

6.7. The depth of a layer of rock can be determined by generating sound waves at the surface of the Earth. Microphones are spread out over the surface to detect sound waves reflected from layers of rock beneath the surface. Table 6.18 shows the distance, d, of each microphone from the point at which the sound was generated and the time delay, t, before the reflected sound wave was detected.

The relationship between t and d is $t^2 = d^2/v^2 + 4h^2/v^2$, where v is the velocity of the wave and h is the depth of the layer of rock.

(i) What would you plot to obtain a straight line?

(ii) Perform a weighted fit to the data to find the slope and intercept of the graph and the uncertainties in these quantities.

(iii) Using the values determined in part (ii), calculate v and h.

Table 6.20 Values of absorbance for various concentrations of $[Cu(H_2O)_6]^{2+}$

$[Cu(H_2O)_6]^{2+}$ concentration (mol/L)	Absorbance (arbitrary units)
0.05	0.206
0.10	0.396
0.15	0.572
0.20	0.740
0.25	0.930
0.30	1.143
0.35	1.347

Table 6.21 Absorbance of standards representing equivalent ALP activity

Equivalent ALP activity (IU/L)[a]	Absorbance (arbitrary units)
0	0.000
25	0.043
50	0.110
75	0.129
100	0.200
150	0.280
200	0.410
250	0.500
300	0.590

[a] The unit IU/L represents international units per litre. Here a unit is defined as the amount of enzyme that will convert 1 mmol of substrate per minute, under defined conditions.

6.8. In an experiment to determine the resistivity of graphite, the electrical resistance of a graphite rod, R, was measured as a function of the distance, l, between electrodes attached to the rod. Table 6.19 contains the resistance–length data.

Assume the relationship between R and l, for a material of cross-sectional area A, can be written $R = \rho l / A$, where ρ is the resistivity of the material.

  (i) Plot a graph of resistance versus length using the data in Table 6.19.

  (ii) Given that the rod has a circular cross-section of diameter (1.26 ± 0.01) cm, calculate the cross-sectional area, A, of the rod and the uncertainty in A.

(iii) Fit a line of the form $y = mx + c$ to the data in Table 6.19 and determine m, c and the uncertainty in each.

(iv) Using values determined in (ii) and (iii), calculate ρ and the uncertainty in ρ.

6.9. Table 6.20 contains values of absorbance for solutions of various concentrations of $[Cu(H_2O)_6]^{2+}$ obtained using visible absorption spectrophotometry.

(i) Find the slope and intercept of a straight line fitted to the data in Table 6.20 using unweighted least squares.

(ii) Calculate the uncertainty in the slope and intercept.

(iii) Use the equation of the line to determine the concentration at which the absorbance is 0.825.

6.10. An absorbance versus alkaline phosphatase (ALP) activity calibration graph was used to assist in the determination of a person's liver function. Calibration data are shown in Table 6.21.

The normal range for ALP activity for a male in the age range 19 to 22 years old is 45 to 150 IU/L. A sample derived from the blood of a male in this age range has an absorbance of 0.337. Use the line of best fit to the data in Table 6.21 to determine whether the ALP activity is outside the normal range.

7 Report Writing and Presentations

7.1 Overview

Progress in science and engineering relies not only on discoveries arising from experiments, but also on communication of those discoveries. In this chapter we focus on three modes of communication, because of their importance in college and university science and engineering courses and their prominence in the professional lives of scientists and engineers: reports, posters and oral presentations.

Where possible, a report or presentation should be planned with the background and interests of the audience in mind.[1] For example, a report written for someone with little background in science would differ from a report written for specialists with expert knowledge of the science being reported. Here we consider communicating the aims, methods and findings of experiments where the audience consists chiefly of science and/or engineering students and their instructors.

7.2 A Scientific or Technical Report

There is often a feeling of accomplishment combined with a sense of satisfaction when an experiment has been completed. The hard work is done, and it is time to relax. However, the background to the experiment, its aim, how it was performed and what was discovered may be known to only a few people at best. The laboratory notebook containing details of the experiment is probably only fully intelligible to its owner and therefore unlikely to provide an easily readable account of the experiment. In situations where all aspects of an experiment need to be communicated, a crucial stage of the work remains to be undertaken: the preparation of a written report.

It is unlikely that you would write a report on completion of every laboratory session you attend. In many situations, especially in the early years of study in

---

[1] It is acknowledged that there may be situations in which you are unaware of the background and interests of your audience, making presentation preparation a challenge.

science and engineering, keeping a laboratory notebook may be given more emphasis with respect to documenting and communicating work done in the laboratory. Nevertheless, it is not unusual to be required to write up one or more experiments in the form of a report in the first year of study at university or college.

As you progress through a course in science or engineering, more emphasis is generally placed on the communication of experimental results. Consequently, the assessment of reports is likely to constitute a significant proportion of marks awarded for experimental work you undertake. The communication may take the form of a report to an instructor or supervisor and be based on an extended experiment or a project.

Beyond this, we recognise that writing reports is a core activity practised by scientists and engineers everywhere. A thorough and well-written report enhances the reputation of the author(s) and can be a decisive factor when decisions are made about providing resources so that, for example, the scope of the work can be expanded.

The process of writing a report requires drawing together and organising the elements that make up a report, including:

- the aim of the experiment
- the background to the experiment
- the method and materials used
- the data gathered
- the analysis and discussion of the data
- the conclusions drawn from the data
- the references used to point the reader to similar work or to support what has been written.

The process of writing a report, which requires revisiting all aspects of an experiment, often rewards its author with a deeper understanding of the experiment. For these reasons, we concentrate first on report writing.

Writing good reports, like proficiency in performing experiments, requires practice, and first efforts can usually be improved upon. Actively looking for the strengths and shortcomings of reports written by others is a good way to identify what makes a competent report, but it is no real substitute for the experience gained through writing one.

A report should:

- be complete but concise
- have a logical structure
- be easy to read.

Some readers may have performed an experiment similar to that discussed in your report, and may wish to compare the experimental method you adopted with their own. For other readers, more general questions are likely to come to mind:

- What was the problem studied and what is its significance?
- What conclusions have been drawn, do they seem reasonable and are they supported by the data?

In addition to addressing these questions, many details must be considered in a report, from describing the background to the work to detailing the equipment used. Dealing with these matters requires a well-structured report.

7.2.1   Structure of a Report

The layout of the report impacts on its clarity, and it is usual for a report to be divided into the following sections.[2]

  (i) Title
  (ii) Abstract
  (iii) Introduction
  (iv) Background Theory
  (v) Materials and Methods
  (vi) Results
  (vii) Discussion
(viii) Conclusion
  (ix) Acknowledgements
  (x) Appendices
  (xi) References

There is no requirement that a report be written in the order given by these section headings. To give the writing of the report impetus, it is sometimes easier to begin with the Materials and Methods section. This section is usually straightforward to write, as many of the details are in the laboratory notebook used to document the experiment.

We will consider each section in turn, but first we begin by considering the use of English in report writing.

7.2.2   Use of English

It is unlikely anyone reads a scientific or technical report primarily to assess the correctness of the English. Nevertheless, if the use of English is poor, a reader may be forced to re-read passages several times to uncover what the writer is trying to say. In extreme cases the reader may give up in frustration.

There is flexibility in how the details of an experiment are reported. It is quite acceptable to write in the first person when describing an experiment. For example:

---

[2] Where a report is short, it is reasonable to combine two or more sections under one heading. As examples, the Introduction may include background theory relevant to the experiment, or Results and Discussion may be combined into a single section. Some sections, such as Acknowledgements or Appendices, may not be applicable and may therefore be absent from the report.

We measured the gas pressure every 30 s. This is certainly easy to read. However, writing in the first person places the emphasis on the person or persons who carried out the activity, rather than on the activity itself.

A more commonly used style in science and engineering is to write in the third person, with the emphasis placed on what was done rather than on who did it. As an example: *The gas pressure was measured every 30 s.* Writing in this manner might leave the reader wondering who did the things described in the report, especially if the report has more than one author. However, this detail is usually of little long-term interest or importance to the reader.

Several other points can be made which should help in the task of report writing.

Choice of Tense

The report is generally written in the past tense when giving details of what was done during the experiment. Occasional use is made of the present tense, especially when giving details of the background to the work or when inferring relationships from the data. For example:

Measurements were made of the length of the copper rod as a function of temperature. The graph of the data shown in Figure 1 **indicates** *that the increase in length of the rod was directly proportional to the temperature rise.*

Sentence Length

Keep sentences short, especially when the content is highly technical or specialised vocabulary is used. The sentence

The decay of photocurrent in a sintered cadmium sulfide photoconductor was measured by illuminating the photoconductor with monochromatic light of wavelength 585 nm for a period of 60 ms, then converting the decaying photocurrent to a voltage using an operational amplifier whose output was connected to a PC-based data acquisition system.

is easier to absorb if it is written as

A cadmium sulfide photoconductor was illuminated with monochromatic light of wavelength 585 nm for a period of 60 ms. The resulting photocurrent was converted to a voltage using an operational amplifier. The decay of the photocurrent was measured by sampling the output of the amplifier using a PC-based data acquisition system.

Acronyms

It is possible that the analysis technique, instrument or device used during the experiment is commonly referred to by an acronym, such as AFM for Atomic Force Microscope or LVDT for Linear Variable Differential Transformer. Though such acronyms may be familiar to some readers of the report, to others they will be new and incomprehensible, which in turn will affect the readability of the whole report. Each acronym should be explained the first time it is used in a report.

7.2.3   Sections of a Report

We will now consider the sections that make up a typical report.

Title and Author

The title of the report should be brief (say between 5 and 15 words), informative and followed immediately by the author's name and their affiliation. Consider the following two titles:

A study of the insulating properties of some materials

Petra Jones, University of the Southern States

and

A comparison of the thermal insulating properties of Styrofoam and fibreglass

Petra Jones, University of the Southern States

The first title is too vague. In the second we are made aware that it is the *thermal* insulating properties that are compared (as opposed to, say, the electrical or sound insulating properties), and specific mention is made of the materials being compared.

Abstract

This is an overview of the experiment and its findings. It should be brief (typically 50 to 200 words) and should avoid the detail that the reader will encounter in later sections. The goal is to get straight to the heart of the matter by communicating what was done, why it is significant and what the major findings are. Writing a good abstract can be quite challenging, and often requires several revisions. Consequently, many people prefer to draw up a plan of the whole report, write a draft of the report and then return to the abstract later.

Consider two alternative abstracts. Abstract A reads:

The cathodoluminescence of a ceramic material is discussed in this report. Cathodoluminescence is the emission of light from a material when it is struck with fast moving electrons. The light emitted from the ceramic was analysed, permitting the identification of a compound formed at the surface. It is possible to relate the existence of the compound to the difficulties some workers have found in making good electrical connections to the ceramic.

Abstract B reads:

Advances in the applications of the superconducting ceramic $YBa_2Cu_3O_{7-\delta}$ have been constrained, partly because of the difficulty of making electrical connections to this material. Cathodoluminescence, used to analyse the surface of the ceramic, revealed that barium carbonate forms at the surface when the ceramic is exposed to air. The barium carbonate forms an electrically insulating layer detrimental to the formation of good electrical connections.

Abstract A focuses on the analysis technique used in the investigation (cathodoluminescence) rather than on the main purpose of the experiment, which was to find out *why* it is difficult to make good electrical connections to the ceramic material. The abstract goes into details of the technique: *Cathodoluminescence is the emission of...* It is important these details be given, but it is preferable that they appear in the Materials and Methods section of the report. Finally, there is a vague statement to the effect that a compound discovered using the technique can help to explain the electrical connection problem. If the author has identified the compound, and the likely relationship between the compound and the problem of making good electrical connections, then this information should be stated concisely in the abstract.

Abstract B is an improvement on abstract A. It states:

- the reason for the investigation
- the type of ceramic material studied
- the technique used
- what that technique revealed
- how the results can be used to explain why electrical connections are difficult to make to the ceramic material.

Abstract B could be improved by including a key quantitative finding, such as the thickness of the layer of barium carbonate.

Introduction

The abstract of the report acts as a summary for the reader. The next stage is to describe the background to the experiment and the goals of the experiment or set of experiments being reported. A reader will 'switch off' if too much detail is given and be confused if there is too little. If the work has followed on from that done by someone else, then that work should be referred to in a manner that will allow the reader to track down details of that work (see the later discussion of the References section).

The introduction should include a description of the background to the work, and outline the specific problem being investigated. The length of the introduction depends on the type of report, but it is unlikely that it would exceed 20% of the whole report. A short report (say 800–1000 words) might have an introduction of about half a page. A good introduction leaves the reader with the sense that the author has a firm grasp of the science or engineering underlying the experiment being reported.

An example is given below of an introduction to a short report. Note how it sets the scene, gives some background to the work and includes references which can be consulted by a reader who wishes to know more.

An electrochromic film changes colour when a voltage is applied across the film. When deposited on glass used for windows, such films can be used to control the amount of light entering a room. Consequently, the films have potential for energy efficiency gains when used in industrial or domestic environments. Established methods for depositing the films include evaporation (Durey et al., 2000) and sputtering (Playfair et al., 2002). A third and possibly more cost-effective method of depositing electrochromic films is to use the sol-gel technique (Rowley et al., 2005). The method requires inexpensive equipment and can be scaled up to provide large area coatings of commercial viability. The technique does have some disadvantages, however. Retention of carbon within films may have a detrimental effect on their optical properties (Harbottle, 2006). This report deals with the deposition of sol-gel films and compares the quality of the films with those prepared using more established techniques. The report focuses on the conditions necessary to produce good films and discusses what improvements are required before the preparation of electrochromic films by the sol-gel deposition technique can be made fully viable.

At the core of many experiments is a hypothesis that is being tested. For example, in an experiment in which the tensile strength of materials is compared, the hypothesis may be expressed as 'the tensile strength of material A is greater than that of material B'. A suitable location for the hypothesis is near the end of the introduction.

Background Theory

If the physical principles underpinning the experiment are well understood and can be summarised by one or two equations, then these may be included in the introduction, with one or more references to sources of further information. The equations may then be referred to, and applied, in the Results section of the report. If the theory is likely to be unfamiliar to the reader, or a short derivation is required, then a separate Theory section can be included within the report.

Materials and Methods

This is a description of how the experiment was performed and how the materials, samples and/or components were used. All important details need to be included in this section of the report. Carefully drawn and labelled diagrams are useful to a reader wishing to visualise the experimental set-up employed.

If a standard experimental technique or protocol has been used, it should be described in a few words. Alternatively, a reference should be given to where full details of the technique can be found.

If you have devised the experimental method yourself or modified an existing method, then include sufficient details so that the experiment can be repeated by someone else. However, this does not mean that the Method section of the report should be presented as a series of instructions such as those found in a laboratory manual. For example, you should avoid writing in a report:

(1) *Connect a 1000 Ω resistor into the circuit as shown in Figure 1.*
(2) *Measure the voltage across the resistor with a digital voltmeter.*

Instead, write:

A 1000 Ω resistor was connected to the circuit as shown in Figure 1. The voltage across the resistor was measured with a digital voltmeter.

Results

It is usually not necessary or desirable to include all the data obtained in the experiment in this section of the report. Doing so may overwhelm the reader with table after table of data. Sufficient representative data should be included so that any discussion that follows or conclusions drawn can be seen to be well supported by the data. Graphs are an excellent way to present large quantities of data, and are likely to be examined before, and more carefully than, data in tabular form.

A graph containing several sets of data aids the reader when data comparison is required. This is preferred to the option of having separate but similar graphs spread over several pages, requiring the reader to switch back and forth between pages to carry out the comparison.

Tables are very useful when summarising values that emerge from the analysis of data. For example, if least-squares fitting has been performed on several sets of x–y data, a table is a compact way to present the parameter estimates, for example of the slope and intercept of the best line and their uncertainties.

Where calculations are presented, attention must be paid to the impact that errors in the 'raw' data have on the calculated values. The subsequent discussion and conclusion will have more credibility if sources of error have been identified and quantified through the inclusion of a statement of the uncertainty in values obtained.

Discussion

The Discussion section deals with the interpretation of the results that have been presented. An experiment is likely to contain many details, both major and minor. Unless the discussion focuses on the important points, a reader is likely to become lost in a mass of unnecessary detail. Where limitations have been identified in the experimental method, these should be discussed. If data from the experiment do not lend strong support to a particular idea or hypothesis at the core of the experiment, then this should be acknowledged. At a minimum, the experiment should provide for a better insight into the problem being studied, and indicate other means of approaching the problem.

Even if the experimental method used could have been improved, we must take care not to be too dismissive of data that were obtained in an experiment. Perhaps better equipment *could* have been used, or more time *could* have been spent collecting data. The question is, what can be usefully said with the data that were gathered, despite any shortcomings?

Conclusion

Here we refer back to the purpose of the experiment. What was the aim of the experiment and was it achieved? If others have undertaken a similar investigation, then it is usual to include a comparison of findings.

It is possible that the value of a quantity has been determined through your experiment which is sufficiently well known that it appears in a data book or textbook. In such a case, a comparison of values should be included along with a reference to the source of the published value. For example, as part of a conclusion to an experiment in which the surface tension of a liquid was investigated, we might write:

An analysis of the data obtained in this work gives a value of $(5.8 \pm 0.2) \times 10^{-2}$ N/m for the surface tension of glycerol at 20 °C. This value compares with that published elsewhere of 6.3×10^{-2} N/m (Johnston, 2017).

Acknowledgements

Experiments in science and engineering are generally collaborative activities. There are situations in which we rely on someone for instruction in the operation of an instrument, for preparation of a sample or for some other form of technical assistance. Perhaps you have discussed the data with someone to clarify your ideas and that person has assisted in improving your understanding, and hence the report. Those who have contributed to the work deserve to be acknowledged. If no mention is made of their help, it would not be surprising to find them less than enthusiastic the next time they are called on for assistance.

Keep acknowledgements brief; for example:

The author gratefully acknowledges Ms J. Sutherland for assisting with sample preparation.

References

References are an important part of a report. They point the reader to relevant information including the background to the work, details of experimental techniques adopted by others who have carried out similar studies, and results obtained by other experimenters. If the experiment is in an area where there have been many previous publications, it will not be possible or desirable to include references to all those publications in the report. Selectivity is the key. If a reference is included to provide general background to the experiment, then a recent book or review article which offers an overview of the subject area is a good option. Such a book or article is likely to have other references which the reader can seek out.

A widespread method of citing references is to include the surname of the author or authors and the year of publication at an appropriate point in the text. For example:

The tunnelling of electrons in semiconductors was first reported in the late 1950s (Esaki, 1958).

The references appearing the References section of the report are given in a standard form.[3] For example:

Esaki, L. (1958) 'New Phenomenon in Narrow Germanium p−n Junctions', Physical Review, 109, pp. 603−604.

It includes

- the name(s) of the author(s)
- the year of publication
- the title of the article
- the journal title
- the volume number, (in this example, the volume number is 109)
- the page numbers.

Another method of referencing is to add a superscript number close to the point at which the reference is relevant. For example:

The tunnelling of electrons in semiconductors was first reported in the late 1950s.[1]

A consecutive list of numbers appears in the References section of the report. Adjacent to each number is the appropriate reference. For example:

[1] *Esaki L 1958 New Phenomenon in Narrow Germanium p−n Junctions Phys. Rev. 109: 603−604.*

An advantage of the first approach, in which the name of the author appears in the citation, over numbering the references is that if another reference needs to be added to the report, it can be inserted without requiring that the remaining references be renumbered.

As well as papers that appear in journals, it is likely that you will want to include references to other sources of information.[4] Useful information may be found, for example:

- in books, including electronic books (e-books)
- on the Internet
- in publications by national or international organisations
- in the proceedings of science or engineering conferences.

---

[3] This is a popular referencing style, known as the Harvard referencing style. There are other, similar, styles in use.

[4] See Pears and Shields (2016) for assistance with referencing, or go to websites such as http://www.citethisforme.com/

Including trustworthy references enhances the authority of a report. For example, including a reference to an article which has been through a refereeing process in which other scientists have reviewed the article (a requirement of most science and engineering journals) is regarded much more highly than those found, for example, on a general website where the correctness of the content cannot be assured.

Appendices

In a report, some material may need to be included which would affect its readability, such as a lengthy derivation of an equation. If no adequate reference to the derivation can be found, the derivation can be included in an appendix. The listing of a computer program written to assist in the analysis of data or an electronic circuit diagram are other examples of items often placed in an appendix.

7.2.4   Preparation Aids

Figure 7.1 illustrates an approach to writing a report, consisting of four stages.

(i) **The plan of the report**. A good plan is one which eases the writing process. Section 7.2.5 describes one approach to planning a report, which involves creating a map of the report.

Figure 7.1 Stages of report writing.

(ii) **The first draft.** This is often written quickly with the goal of setting down essential information in each section, and with less consideration given to matters such as correct punctuation or consistent use of tense. Typically, the first draft contains incomplete sections which are returned to later.

(iii) **Reviewing what has been written.** It is valuable to ask someone who might be able to offer you some feedback or suggestions for improvement to review the first draft. If you do the review yourself, then allowing a day or two between writing the draft and reviewing it, if possible, will assist you to examine the draft with a 'fresh eye' and detect mistakes or omissions that you would have otherwise overlooked.

(iv) **Revising the draft.** The draft is revised following the review. It is common for the new draft to be reviewed before the content is settled and the report made available to others.

When you have completed the report, it can be helpful to read it aloud. This will assist in uncovering small mistakes, such as missing words or poor grammar, that may have been there since the first draft but were not discovered during the review and revision process.

7.2.5   The Plan: A Map to Aid Report Preparation

A one-page diagram or map containing the essential features of a report can be a useful aid when planning a report. The map[5] can:

- assist in organising the contents of the report
- trigger ideas about what else to include
- emphasise relationships between the contents of each section of the report
- alert the author to details that have been overlooked or need more attention
- include key facts and figures to be included in the report.

As report writing proceeds, the map is revisited and modified if necessary. Figure 7.2 shows part of a map created to assist in writing a report on an experiment carried out on a helical spring. The map consists of shapes (often hand-drawn circles or rectangles) containing text, symbols and equations. The shapes are linked with lines indicating a relationship between each shape. The part of the map shown in Figure 7.2 is a summary of the Materials and Methods section of the report.

---

[5] This is sometimes referred to as a 'mind map' or a 'cognitive map'.

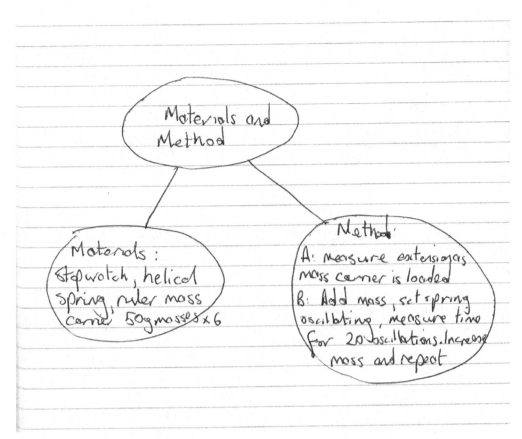

Materials and Method

Materials :
Stopwatch, helical spring, ruler mass carrier 50g masses × 6

Method :
A: Measure extensions as mass carrier is loaded
B: Add mass, set spring oscillating, measure time for 20 oscillations. Increase mass and repeat

Figure 7.2 Section of a map created to assist in writing a report on the properties of a helical spring.

Computer-generated maps can be created in several ways, including using the SmartArt tool in Microsoft Word.[6] However, a hand-drawn map can generally be constructed more quickly, especially if equations are to be included on the map. A map can be conveniently located in the laboratory notebook used to document the experiment (and might cover two pages, if the map is large). Figure 7.3 shows a map used in the preparation of a report on an experiment which investigated the electrical resistivity of graphite.

It is convenient to place the theme of the report – expressed, say, by a provisional title – near the centre of the map. Lines are drawn radiating from the central circle to others containing many of the headings appearing in Section 7.2.1. Within most circles is some key content to be included and expanded upon in the report. Some

---

[6] https://mindmapsunleashed.com/learn-to-create-a-mind-map-in-word-heres-how

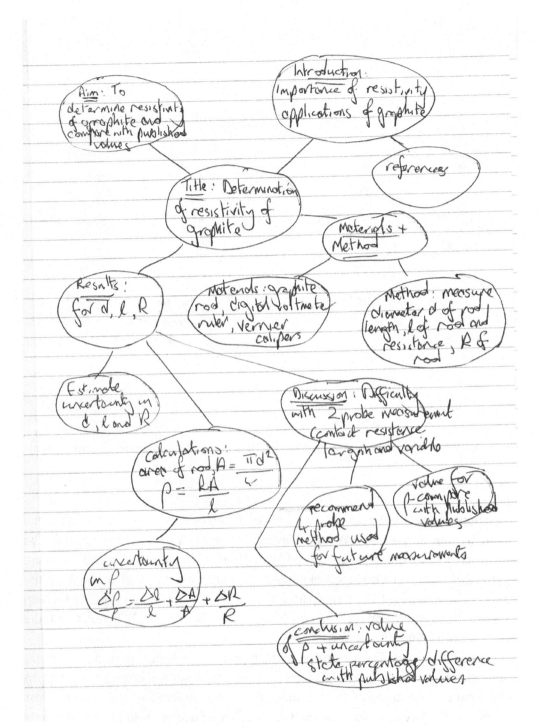

Figure 7.3 Map of a report on an experiment to measure the resistivity of graphite.

circles, such as the one representing the discussion, are likely to contain several ideas that might be included in the report. In such a case, further sub-branching lines are incorporated, leading to other circles containing those ideas. More branches and circles are added as the map grows.

As a map is an intermediate step in preparing a report, there is little need to be concerned about its neatness or elegance. If the map is meaningful to its creator, it will be of value as report writing proceeds. Reports have quite a rigid structure and are prepared for distribution to, and examination by, others. By contrast, a map is primarily an aid to the author and unlikely to be inspected by anyone else.

Having the whole report represented as a map on a single sheet of paper can bring emphasis to essential features of the report, such as what should be in each section and how sections link together. I reflect that, as I have become more experienced as an experimenter, I use maps regularly when preparing reports (as well as when preparing posters and oral presentations). However, not everyone finds maps useful when preparing a report. Some authors prefer to begin their report by simply typing the section headings appearing in Section 7.2.1 and then populating those sections with words, figures, charts and tables.

Exercise A

An experiment is carried out in which two methods for measuring the spring constant of a helical spring are compared. The report based on the experiment, given next, is laid out in the manner described in this chapter. Sections of the report can be improved. Read the report, then focus on each section and address the following questions:

(i)   What is good about the section?
(ii) How could the section be improved?

Two Methods Compared for Determining the Spring Constant of a Helical Spring

D DUREY

Abstract

Springs are widely used in domestic and industrial situations. The spring constant is an important characteristic of a spring. In this study the spring constant of a helical spring was established by two methods, one static and the other dynamic. The degree of mutual consistency for the spring constant determined using both methods was good, with a value of $(13.07 \pm 0.04)\,\text{N/m}$ obtained using the static method and a value of $(13.42 \pm 0.08)\,\text{N/m}$ for the dynamic method. Owing to the ease of the method, and the smaller measurement uncertainty it yields, the static method is preferred for determining the spring constant.

Introduction

Springs have many applications, including isolating buildings from the effect of earthquakes, in automobile suspensions to improve the comfort of the ride for passengers and in spring balances used to measure forces (Bankes et al., 1998; Parvin and Ma, 2001; Yamada, 2007). An important characteristic of a spring is its spring constant. The aim of this study was to compare two methods for determining the spring constant of a small helical spring.

The spring constant, k, can be found by applying a force, F_{app}, to a spring and measuring the extension, x, that the force produces. The relationship between F_{app} and x is (Walker et al., 2014)

$$F_{app} = kx. \tag{1}$$

Rearranging equation (1) gives

$$x = \frac{F_{app}}{k}. \tag{2}$$

The spring constant may be found by measuring x as F_{app} is varied.

An alternative approach to determining the spring constant is to analyse the oscillation of a mass attached to the end of the spring. When the mass on the end of a spring undergoes simple harmonic motion, the period T of the motion is given by (Serway and Vuille, 2015)

$$T = 2\pi\sqrt{\frac{m}{k}}, \tag{3}$$

where m is the mass attached to the spring.

Squaring both sides of equation (3) gives

$$T^2 = 4\pi^2\frac{m}{k}. \tag{4}$$

By varying m and measuring T, equation (4) can be used to determine the spring constant.

Materials and Methods

Method A considers the extension produced by an applied force and Method B the period of oscillation as a function of mass added to the spring. Method A and Method B were applied to the same spring.

Method A

A small spring of length 20 cm was suspended vertically from a rigid point. A mass holder was attached to the free end of the spring and a ruler positioned adjacent to the spring, as shown in Figure 1, to allow the extension of the spring to be measured. A pointer was used to indicate the position of the end of the spring. Standard 50 g slotted masses were added, one at a time, to the holder and the new position of the pointer recorded after each mass was added. This allowed the extension of the spring to be established for a range of force applied to the spring.

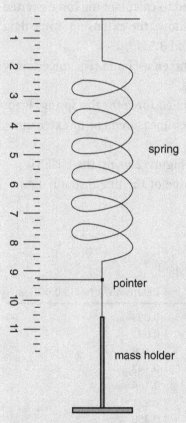

Figure 1 Experimental arrangement.

Method B

A standard 50 g mass was added to the mass holder in Figure 1 and allowed to move down to a new equilibrium position. The mass holder was pulled down one centimetre and then released. The time was recorded for 20 complete oscillations of the mass holder. To check for repeatability of measured values, and to improve precision of the measurement, the timing of 20 oscillations was repeated twice.

Further 50 g masses were attached to the holder, one at a time, and the procedure repeated.

Results

The spring constant obtained using Method A is written k_A and that obtained using Method B is written k_B.

Method A

The relationship $F = mg$, with $g = 9.81$ m/s, was used to calculate the force exerted by the loaded mass holder on the spring. Table 1 shows the extension–force data for applied forces between approximately 0.5 N and 3.5 N.

The uncertainty in reading the metre ruler was taken as 1 mm. The uncertainty in the standard masses was taken to be negligible.

Figure 2 shows a graph of extension versus applied force for the spring. Error bars which would represent the uncertainty in each measurement of extension are too small to plot on this graph.

A straight line was fitted to the data shown in Figure 2 using the LINEST function in Excel.[7] Applying the tool gives the slope of the line through data

Table 1 Extension–force data for helical spring

Mass added, m (kg)	Applied force, F_{app} (N)	Extension, x (m) (± 0.001) m
0.050	0.491	0.033
0.100	0.981	0.071
0.150	1.472	0.108
0.200	1.962	0.147
0.250	2.453	0.184
0.300	2.943	0.221
0.350	3.434	0.258

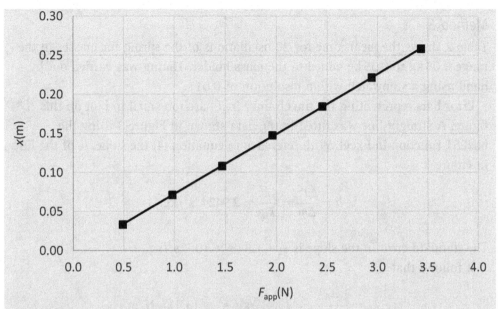

Figure 2 Plot of extension–force data shown in Table 1.

and the standard error in the slope. With reference to equation (1) the slope, b, of the line in Figure 2 is

$$b = \frac{\Delta x}{\Delta F_{app}} = \frac{1}{k_A} = 0.07653 \text{ m/N}.$$

The standard error in the slope is $s_b = 2.46 \times 10^{-4}$ m/N.
   It follows that

$$k_A = \frac{1}{b} = 13.068 \text{ N/m}.$$

The standard error in k_A is

$$s_{k_A} = \left|\frac{\partial k_A}{\partial b}\right| s_b = \frac{1}{b^2} s_b = \frac{1}{13.068^2} \times 2.46 \times 10^{-4} \text{ N/m} = 0.0421 \text{ N/m}.$$

So, $k_A = (13.07 \pm 0.04)$ N/m.

---

[7] Note: Excel's LINEST function is discussed in Section 8.3.2.

Method B

Table 2 shows the mean time for 20 oscillations of the spring for masses in the range 0.05 kg to 0.35 kg added to the mass holder. Timing was carried out by hand using a stopwatch with a resolution of 0.01 s.

Error bars representing the uncertainty in T^2 are too small to plot on this figure. A straight line was fitted to the data shown in Figure 3 using the LINEST function in Excel. With reference to equation (4) the slope, b, of the line in Figure 3 is

$$b = \frac{\Delta T^2}{\Delta m} = \frac{4\pi^2}{k_B} = 2.9424 \ \text{s}^2/\text{kg}.$$

The standard error in the slope is $s_b = 1.848 \times 10^{-2}\,\text{s}^2/\text{kg}$.

It follows that

$$k_B = \frac{4\pi^2}{2.9424} = 13.417 \ \text{kg/s}^2 \equiv 13.417 \ \text{N/m}.$$

The standard error in k_B is

$$s_{k_B} = \left|\frac{\partial k_B}{\partial b}\right| s_b = \frac{4\pi^2}{b^2} s_b = \frac{4\pi^2}{2.943^2} \times 1.750 \times 10^{-2}\text{N/m} = 0.080 \ \text{N/m}.$$

So, $k_B = (13.42 \pm 0.08)\,\text{N/m}$.

Table 2 Variation of time for 20 oscillations of the spring as a function of mass. Also shown are the calculated values for the period, (period)^2 and the uncertainty in each of these.

Mass added, m (kg)	Mean time for 20 oscillations (s)	Uncertainty in mean time (s)	Period, T (s)	Uncertainty in T (s)	Period^2, T^2 (s^2)	Uncertainty in T^2 (s^2)
0.05	11.38	0.16	0.569	0.0080	0.324	0.009
0.10	13.79	0.19	0.690	0.0095	0.476	0.013
0.15	15.81	0.16	0.791	0.0080	0.626	0.013
0.20	17.51	0.25	0.876	0.0125	0.767	0.022
0.25	19.23	0.10	0.962	0.0050	0.925	0.010
0.30	20.60	0.12	1.030	0.0060	1.061	0.012
0.35	21.97	0.06	1.099	0.0030	1.208	0.007

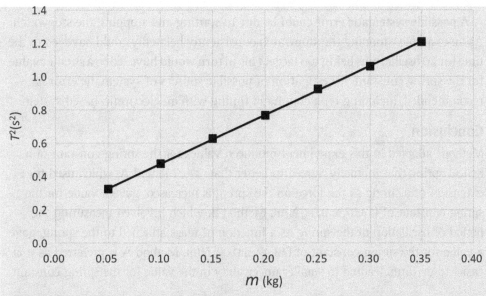

Figure 3 Plot of T^2 versus m data given in Table 2.

Discussion

Equations 2 and 4 predict linear relationships between the dependent and independent variables. The prediction is supported by the graphs in Figure 2 and Figure 3.

Being a static measurement, the extension versus applied force experiment was easier to perform than the dynamic experiment in which the time for 20 oscillations was measured. This is possibly the main reason for the smaller value for the uncertainty in the spring constant arising from the static experiment. More specifically, the standard error in the spring constant using Method B is twice that of Method A. For small mass added to the spring, the period of the oscillation was small, requiring careful concentration to count 20 oscillations. Further, hand timing 20 oscillations required synchronising the starting and stopping of the stopwatch with the release of the mass and the completion of 20 oscillations, respectively. It was difficult to judge when the last of the 20 oscillations had been completed.

The values for the spring constant using the two methods differ from each other by 2.7%. This difference cannot be attributed solely to the random errors arising using methods A and B; the standard errors in k estimated using each method are quite small, such that the 95% confidence intervals do not overlap. This implies that the true values for the spring constant obtained using each method are not the same. This suggests that there are systematic errors that arise when employing Method A and/or B that have not been accounted for, or that equations 1 and/or 3 do not completely account for the behaviour of the spring.

A possible systematic error could be due to starting and stopping the stopwatch. As an example, stopping the stopwatch consistently belatedly could have led to the time for 20 oscillations being too large. This in turn would have led to a smaller value for the spring constant. Investigation of possible sources of systematic error is recommended, including replacing hand timing with an electronic-based system.

Conclusion

Methods adopted in this experiment produced values for the spring constant of a helical spring that mutually agreed to better than 3%. Method A, which used the extension of a spring as the force on the spring is increased, gave a value for the spring constant of (13.07 ± 0.04) N/m. Method B, which involved measuring the period of oscillation of the spring as a function of mass attached to the spring, gave a value for the spring constant of (13.42 ± 0.08) N/m. Method A is preferred as it is easier to perform, leading to smaller uncertainty in the value for the spring constant.

REFERENCES

Bankes, M. J. K., Crossman, J. E. and Emery, R. J. H. (1998) 'A standard method of shoulder strength measurement for the Constant score with a spring balance', *Journal of Shoulder and Elbow Surgery*, 7(2), pp. 116–121.

Parvin, A. and Ma, Z. (2001) 'The use of helical spring and fluid damper isolation systems for bridge structures subjected to vertical ground acceleration', *Electronic Journal of Structural Engineering*, 1(2), pp. 98–110.

Serway, R. A. and Vuille, C. (2015) *College Physics*. 10th edn. Connecticut: Cengage Learning.

Walker, J. *Fundamentals of Physics*. 10th edn. New John Wiley and Sons: York.

Yamada, Y. (2007) *Materials for Springs*. Berlin: Springer-Verlag.

7.3   Posters

Posters are widely used for communicating the aims, methods and findings of experiments in colleges and universities, as well as in professional settings such as meetings, conferences and other gatherings of scientists and engineers. A poster offers scope for creativity and imagination while at the same time being a clear, concise and accurate description of an experiment, its findings and conclusions.

Posters share common features with reports but are more succinct descriptions of experiments and their findings. Well-crafted posters are clear and concise, display technical proficiency and demonstrate understanding of the science or technology being presented. Posters offer the opportunity for creativity on the part of the author(s). This could come in the form of an eye-catching layout or the imaginative use of images.

7.3.1   Poster Preparation

Posters comprise words, tables, diagrams and charts. They are often assembled with the aid of a computer package such as Microsoft PowerPoint, Microsoft Publisher or SciGen Technologies Poster Genius. Poster templates suitable for communicating experiments, many which are free, may be found on the Internet and are a useful starting point for creating a poster.[8] Such templates offer support with the technical aspects of poster creation and assist by suggesting layout options, colour schemes and choice of font and font size. Adapting an existing template is often an attractive alternative to creating one from 'scratch'.

Posters usually adopt similar section headings to those found in a report. These assist readers to navigate the poster, or perhaps to jump to the section which most interests them, such as the conclusion. Much of the advice on report writing offered in previous sections of this chapter also applies to posters. However, a poster is *not* a report 'cut and pasted' into a poster template. Such a poster would be word-intensive and visually unappealing. Each section of a poster should be brief and include only essential details.

Once complete, posters are printed and sometimes laminated for protection.[9] Poster sizes may be metric or non-metric and of any size. Typical poster dimensions are 841 mm × 1189 mm (metric size A0), 594 mm × 841 mm (metric size A1), 36 inches × 48 inches and 36 inches × 60 inches. Such sizeable posters allow for large font sizes, diagrams and charts, permitting a poster to be read from a typical viewing distance of about 2 m.

Posters are generally displayed where people can walk up to them and stop to ask questions of the author. This allows an author to:

- guide a viewer through the poster, pointing out key features
- expand on details given in the poster
- clarify aspects of the poster.

Once the draft of the poster is complete, it is useful for its author to ask: 'If I were viewing the poster for the first time, what would I want to know?' This can help you revise the contents of the poster. Another, and perhaps superior, method is to approach someone whose insight you trust and have them ask you questions about the poster.

We now offer an example of a poster along with an account of its layout and contents.

7.3.2   Poster Example

There is no rigid recipe for devising a technically accurate, visually attractive poster.[10] In Figure 7.4 we present a poster which contains elements of a good poster.

---

[8] Free templates can be found at https://www.posterpresentations.com/free-poster-templates.html

[9] In some circumstances a poster may remain in electronic form and be shown to a large audience using a data projector.

[10] Advice on poster preparation can be found at https://guides.nyu.edu/posters

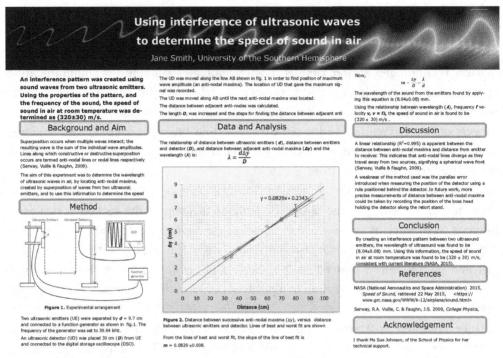

Figure 7.4 Example poster.

The poster shown in Figure 7.4, which was prepared using Microsoft Publisher, was created to communicate the findings of an ultrasonic waves experiment. The layout adopted consists of three columns with a graph as the centrepiece. The black text on a white background gives maximum contrast, an aid to reading the text from a distance. Approximately 600 words appear on the poster. The poster in Figure 7.4 consists of:

- title describing the experiment followed by the name of the author and her affiliation
- summary of the experiment and the key finding (the speed of sound in air). To give the finding prominence, the summary is bolded and in larger font size than the text in the sections that follow.
- Background and Aim section, which includes information that assists in setting the scene for the experiment and concludes with the aim of the experiment.
- Method section, which includes a diagram of the experimental set-up. Although a line diagram is used here, a photograph of the apparatus could have been included instead. Sufficient detail of the method has been incorporated to allow someone else to repeat the experiment.
- Data and Analysis section, which begins with the equation describing the relationship between the variables studied in the experiment. As space is at a premium, no tabulated data are included on the poster. Instead, a graph is used to display the data, along with a line of best fit and two other lines that allow the uncertainty in the slope to be calculated.

- Discussion section, which points to the significance of the data, and draws attention to a weakness in the method used.
- Conclusion section, which restates the main finding of the experiment and indicates that the value obtained for the speed of sound in air is in keeping with that published elsewhere.

Two references and an acknowledgement complete the poster.

Exercise B

Consider the poster in Figure 7.5. Describe three strengths of the poster and suggest three improvements that could be made to the poster.

The oral presentation of the findings of experiments is another common mode of communication in science and engineering. In the next section we consider the oral communication of experiments.

7.4   Oral Presentations

An oral presentation is an opportunity to give a personalised account of an experiment, place it in context, guide an audience through aspects of the experiment and highlight important features and findings. It also allows for questions and answers from members of the audience so that they can seek clarification or further explanation of points raised during the presentation.

Oral presentations are often short, with 10 to 15 minutes being typical. Such a short duration requires that presentations be carefully planned to reduce the risk of running overtime.

Standing up and presenting work to an audience can be an uncomfortable experience, and it is an unusually self-confident person who can do this without feeling nervous. Being thoroughly familiar with the experiment to be presented, and its background, purpose and findings, will lessen presentation anxiety.

Preparation

If the oral presentation is based on a written report, then methodically re-examining the report contents is an excellent starting point for preparing the presentation. It is likely, for example, that the report contains figures and graphs that can be adapted for inclusion in the presentation. Short notes can be made of sections of the report and kept nearby during the presentation. Having notes to hand is reassuring[11] if you lose your way, and will help you feel composed during the presentation.

---

[11] Even if you don't find yourself referring to them.

Measurement of the acceleration due to gravity using a simple pendulum

Introduction All bodies experience a force due to gravity. The effect of the force is most obvious when a body is allowed to fall freely under the influence of gravity. Newton's law of gravitation predicts that the acceleration due to gravity, g, on the Earth's surface depends on the mass of the Earth and the distance of the falling body from the centre of the Earth. As a body close to the Earth's surface falls a small distance compared to the radius of the Earth, g, can be regarded as constant over that distance (Serway and Vuille, 2015).

Aim The aim of this experiment is to establish the value of g near to the surface of the Earth by analysing the motion of a simple pendulum.

Method A small ball was attached to a fixed point by a thread. The ball was allowed to swing freely through a small angle ($\approx 5°$) as shown in Figure 1.

The stop-watch shown in Figure 2 was used to determine the period of the motion, T, of the pendulum. The time for five successive swings was measured to reduce the influence of reaction time on the uncertainty in T.

Theory The relationship between the period, T, of the motion of a simple pendulum and its length, L, is (Walker, 2014):

$$T = 2\pi\,(L/g)^{\frac{1}{2}} \qquad \text{Equation 1}$$

This equation is of the form $y = mx$ with slope, m, given by,

$$m = 2\pi/(g)^{\frac{1}{2}} \qquad \text{Equation 2}$$

Results Table 1 shows the data gathered for the period of the pendulum as a function of the length of the pendulum.

L (m) ± 0.005 m	T (s) ± 0.02 s
0.410	1.32
0.575	1.55
0.700	1.71
0.825	1.84
0.900	1.93

Table 1: Period of pendulum for various thread lengths.

Figure 1: Experimental arrangement

Figure 2: Stopwatch

Analysis

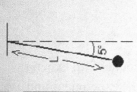

Figure 3: Graph of pendulum period versus (length)$^{\frac{1}{2}}$

Line	slope (s m$^{-\frac{1}{2}}$)	acceleration due to gravity (m s^{-2})
1	2.12	8.78
2	2.00	9.87
3	1.88	11.2

Table 2: Slopes of lines shown in Figure 3.

Figure 3 shows a plot of the period of T versus $L^{\frac{1}{2}}$. Table 2 contains the slopes of the three lines drawn through the data points along with the corresponding values for the acceleration due to gravity found using equation 2.

Discussion The value for g obtained here is consistent with that published elsewhere (Haynes, 2016). The main source of uncertainty is timing the duration of 5 swings of the pendulum. This is caused by the difficulty in synchronising the starting and stopping of the stopwatch with the pendulum motion. To reduce this uncertainty, it is recommended that the number of swings be increased.

Conclusion Based on the data collected in this experiment, the acceleration due to gravity is found to be (9.9 ± 1.2) m s^{-2}

References
Serway and Vuille (2015) *College Physics 10th Edition*, Cengage: Stamford
Walker J (2014) *Fundamentals of Physics 10th Edition*, Wiley: New York
Haynes W M (2016) *CRC Handbook of Chemistry and Physics 97th Edition*, CRC Press, New York

Figure 7.5 Poster describing the determination of the acceleration due to gravity.

If a report of the experiment does not exist, then a map, similar to that described in Section 7.2.5, can assist in structuring the presentation. A written outline can also serve as a memory aid.

An oral presentation typically consists of four parts: introduction, body, conclusion and question and answer session.

Introduction

It is normal to introduce yourself and announce the title of your talk. Next comes an overview of the topic of the experiment. This is an excellent opportunity to capture the attention of the audience, especially those in the audience with little knowledge of what you have been working on. When you are very familiar with an experiment it is easy to plunge into too much detail too early, leaving the audience struggling to grasp the 'bigger picture' – for example: what is the background to the experiment; what was the experiment about; what is its significance? The introduction often concludes with a slide indicating the content of the remainder of the talk. For a presentation of 10 to 15 minutes, 2 to 3 minutes are typically allocated to the introduction.

Body of the Talk

Here details of the methods used, the data gathered, the challenges faced and the analysis undertaken are described. Presenting these using diagrams and graphs is preferred, for example, to slides consisting of tables of data, as these take longer for the audience to assimilate. If is it necessary to include equations, then these should be kept to a minimum; a slide containing many equations, especially if unfamiliar, is difficult to comprehend in the time available. For a presentation of 10 to 15 minutes, 6 to 8 minutes are typically allocated to the body of the talk.

Conclusion

Here the audience is reminded of the context of the experiment and its purpose. The main finding of the experiment is restated and emphasised, along with suggestions for improvements to the experiment or for future work. For a presentation of 10 to 15 minutes, 1 to 2 minutes are typically allocated to the conclusion. As the presentation nears the end you would acknowledge those that had helped you with the experiment. It is customary to thank the audience for their attention.

Question and Answer Session

This can cause the most anxiety, as you have now transferred some control of the presentation to the audience. With a short presentation, you may typically be asked two or three questions. If there is a chance that the whole audience has not heard a question, then repeat it for all to hear. In addition to making everyone aware of the question being asked, repeating the question gives you valuable thinking time. This is also the point at which you can ask for clarification if the question is not clear.

Answering questions about experimental methods used, challenges faced or analysis methods adopted is usually straightforward, as you will have spent time on these during and after the experiment. Try to keep answers short and to the point.

It is possible you will be asked a question which you cannot answer, perhaps because it is beyond the scope of the experiment or touches on something of which you have little experience. Don't be afraid to admit you don't know enough to be able to answer the question; thank the questioner for alerting you to something you need to follow up on.

Visual Aids

It is quite rare for an oral presentation in science or engineering to be delivered without some form of computer-based visual aid such as Microsoft PowerPoint. PowerPoint has been the leading presentation software for more than 20 years and it is likely that any computer you use will have PowerPoint installed. However, there are many other presentation software options available, including Prezi.[12]

A short video can be embedded in a presentation and is useful if you want to present something that is dynamic by nature (for example, the simulation of a chemical reaction or the motion of a satellite).

Delivery

It is better to avoid reading from notes, or word-for-word from a slide that those in the audience can see and read well enough for themselves. While the odd 'err' or 'um' in a presentation is normal, too many are distracting. Smiling will help the audience connect with you, as will showing enthusiasm throughout your presentation.

For a 15-minute presentation, 10 to 15 slides would be normal, so long as the slides do not contain a lot of detail. If there are details that the audience does not need to see,[13] then it is better to remove them than to say, for example, 'please ignore the tables of data in this slide'. Where something can be presented as text or images, choose images, as they tend to be more memorable.

Practice

Practicing a presentation, preferably with someone watching and listening who can give you immediate feedback, leads to an enhanced presentation. If no one is available to give you feedback, then capturing a practice session, for example on a smartphone, will assist you to:

- get the timing right; the audience, and those waiting to present their work after you, will not thank you if you run over time
- improve the sequencing of the presentation and help ensure that you move smoothly from one idea to the next

---

[12] https://prezi.com/

[13] Or cannot see because the details are too small.

- recognise distracting body language, mannerisms or gestures you were unaware of – for example, looking at your notes or the screen rather than directly at the audience, or standing with your hands in your pockets.

Technical Aspects

A well-prepared presentation can be derailed by technical difficulties such as a PowerPoint file that will not load or video clips that refuse to play or that play with no sound. It is not possible to be completely confident that a presentation will go smoothly from a technical perspective. However, loading and quickly stepping through the presentation in the venue before it is due to begin can assist in identifying and addressing most difficulties.

7.5 Comment

The findings of experiments in science and engineering are communicated through reports, posters and oral presentations. It could be argued that without communication of the purpose, results and outcomes of an experiment, the effort expended in performing the experiment has been largely wasted.

A report is a permanent record of work that can be referred to, archived and used by others. When the minute details of an experiment are forgotten, the report endures as an account of what was done, and is the most conspicuous indicator of how well it was done.

Papers published in journals devoted to science or engineering bear close similarity to reports. One difference is that papers are in the public domain and generally accessible to a larger audience than most scientific or technical reports. In large part, the structure of a paper follows that described in Section 7.2.1. Journals do vary regarding the advice they offer to authors preparing a paper. Nevertheless, much of the advice in this chapter on report writing also applies to paper writing.

A poster is a compact way to present the findings of experiments. The emphasis is on the economical use of words while maximising the use of graphs, figures and other images. A poster should be self-explanatory and need no further input from its author. An attractive poster draws an audience, allowing an author located nearby to elaborate on what was done and how it was done, and to answer questions related to the poster.

Delivering an oral presentation about an experiment you have carried out can be quite unnerving, but has the potential for more impact than either a report or a poster. A well-planned presentation, followed by practice, review and timely feedback, will assist you to develop and enhance a skill of lasting value in any future professional role you might assume.

8 Using Excel to Present and Analyse Data

8.1 Overview: Spreadsheets for Data Analysis

So long as there are not too many data to analyse, the techniques that we have considered so far, for example fitting a straight line to x–y data, are fairly easy to apply with the aid a scientific calculator. However, as the amount of data increases, so does the likelihood of a mistake such as entering an incorrect number into the calculator.

Some calculations are unwieldy, such as when a weighted least-squares fit to data is done. Under normal circumstances this requires that a table be drawn up containing many columns, and demands numerous time-consuming calculations.

It would be useful to have a computer-based analysis program into which we could enter data and have calculations performed automatically, such as those required when fitting an equation to that data using least squares. It would be informative to have the option of displaying the results of calculations in graphical as well as numerical form, and for data to be easily added to and modified. In situations where a computer has been used to gather data directly, it would be appealing to transfer data directly to the program for analysis and presentation purposes, without the chore of first recording numbers and then entering them by hand.

Computer-based spreadsheets with these capabilities have been available since the early 1980s. Since that time, spreadsheets have developed to the extent that they have become powerful tools for analysing and presenting scientific and technical data. In this chapter we explore some of the basic features of spreadsheets and show how they may be applied to the analysis of experimental data. Because of its availability, power and longevity, we focus on Excel, a spreadsheet program developed by Microsoft,[1] although other spreadsheet programs have similar features.[2]

---

[1] A useful introduction to Excel is Excel Easy, https://www.excel-easy.com/ . For books that focus on the use of Excel in a scientific or engineering context see Liengme (2015) and Kirkup (2012).

[2] Calc by Apache OpenOffice, and Calc Spreadsheet by LibreOffice are free spreadsheets incorporating many of the same features and functions as Excel. See http://www.openoffice.org/ and https://www.libreoffice.org/

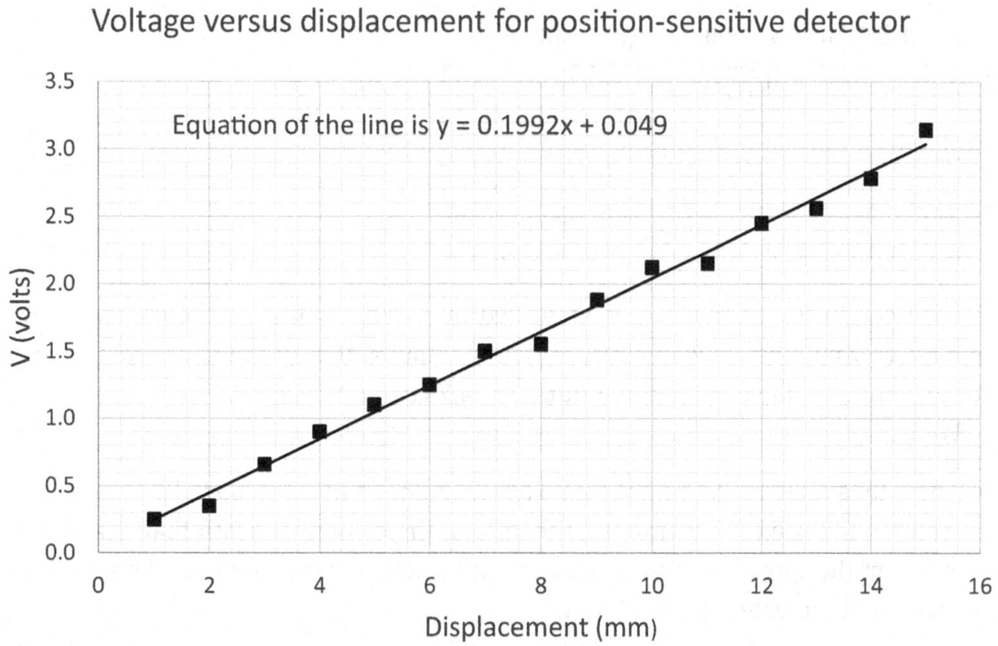

Figure 8.1 Example of an *x–y* graph produced by a spreadsheet.

8.1.1 Illustration of Experimental Data Analysis Using a Spreadsheet

Figure 8.1 illustrates what can be done in a few minutes with a spreadsheet. An experiment has been performed to study the linearity of a position-sensitive detector. The output voltage from the detector has been recorded as a function of the position of a light spot which is focused on the active area of the detector. Data are entered into the spreadsheet in columns and plotted in the form of an *x–y* graph.[3] The spreadsheet has been used to calculate the slope and intercept of the best straight line using equations 6.7 and 6.8, respectively, and to plot this line on the graph.

Whenever data in a spreadsheet are changed (which could be necessary if, for example, a mistake has been made when entering the data), the graph is updated, and linked calculations, such as those that determine the slope and intercept of best line through *x–y* data, are automatically updated.

8.2 Spreadsheet Basics

A spreadsheet consists of a two-dimensional array of boxes, referred to as *cells*,[4] as shown in Sheet 8.1. To indicate clearly that we are dealing with a spreadsheet here, we refer to it as a 'sheet' and show it with a grey background and bordered by two lines.

---

[3] The visual presentation options in Excel are referred to as 'charts'. The *x–y* graph is referred to in Excel as an *x–y* Scatter Chart.

[4] Normally somewhere in the region of 10 columns by 15 rows of cells are displayed on the screen of the computer, but for convenience we show only a small part of the sheet in Sheet 8.1.

Sheet 8.1 Small section of a spreadsheet

	A	B
1		
2		14.751
3		

Each cell in the spreadsheet is identified by a combination of the column letter and the row number. Each cell can contain a number, text, formula or function. So, for example, we have entered the number 14.751 into the cell B2.

We can:

- draw up a table containing the data as we might do in a notebook by giving each column a heading and entering the data in the column beneath that heading
- instruct the spreadsheet to perform an arithmetic operation on the data, such as summing the values in a column.

Where then, is the advantage of a spreadsheet over drawing up a table by hand in a notebook and using a calculator to, for example, sum values in the table? It is this: once the values have been entered, along with formulae which use those values, any subsequent change will cause the result of calculations to be automatically updated. This apparently simple feature is extremely convenient when many interlinked calculations are performed on data.

To illustrate this, Sheet 8.2a shows a spreadsheet containing the values of 10 precision resistors of nominal value 33.0 Ω, whose resistance has been measured with an instrument of resolution 0.1 Ω. To add up all the values of resistance, the formula =SUM(B1:B11) is typed into cell B12.[5] When the Enter key is pressed, the sum of the values in cells B1 to B11 is calculated and the result (in this case 326.7) is 'returned' into cell B12.

The value 33.9 in the highlighted cell, B7, in Sheet 8.2a has been entered incorrectly and is to be replaced by 29.3. Cell B12 shows the sum of the 10 values; as soon as the correct value is entered, the sum is updated from 326.7 to 322.1, as seen in Sheet 8.2b.

8.2.1   Calculations Involving Columns of Data

Once the values from an experiment have been recorded, it is likely that we will want to perform a variety of arithmetic operations on them, such as addition, subtraction and multiplication. Or perhaps we would like the natural logarithm of each value to be calculated.

---

[5] We adopt the convention that anything to be typed into a cell appears in **bold**.

Sheet 8.2a Section of spreadsheet containing an incorrectly entered value

	A	B
1		Resistance (Ω)
2		33.4
3		33.8
4		32.2
5		31.9
6		32.2
7		33.9
8		34.0
9		31.4
10		31.6
11		32.3
12	sum =	326.7

Sheet 8.2b Updated spreadsheet after correcting the value in cell B7

	A	B
1		Resistance (Ω)
2		33.4
3		33.8
4		32.2
5		31.9
6		32.2
7		29.3
8		34.0
9		31.4
10		31.6
11		32.3
12	sum =	322.1

Sheet 8.3 Number of counts recorded for various thicknesses
of absorbing material

	A	B	C
1	Thickness (mm)	Counts	ln(Counts)
2	1	1213	= LN(B2)
3	2	785	

As an example, consider an experiment in which the number of gamma rays that have passed through a material is counted. Sheet 8.3 shows the number of counts recorded (over a period of one minute) as the thickness of the absorbing material is varied. One step in the analysis to determine the capability of the material to absorb the gamma radiation requires that we take the natural logarithm of the recorded counts. This is a tedious process with a hand calculator, especially when there are many values. By contrast, this operation is easy to do on all the values 'simultaneously' using a spreadsheet.

The first step is to have the spreadsheet calculate the natural logarithm of the contents of cell B2. In Excel this is done by making C2 the active cell[6] and typing into it the function which will calculate the logarithm of the contents of B2. To indicate that we are about to enter a formula (and not data or text) into C2, we begin with an equals sign, then follow this with the function, as shown in Sheet 8.3.[7] Note that Excel uses upper-case letters to indicate any mathematical function (e.g. LN(B2), rather than ln(B2)), although we can type in the functions in either upper or lower case.

Once the Enter key is pressed, cell C2 contains the calculated value, as shown in Sheet 8.4.[8]

The next step is to fill a group of cells with a selected formula. This allows the logarithms of the remaining numbers to be calculated. In this example, a FILL command is used to fill C2 through C11 with =LN(B2) through =LN(B11), respectively. To do this:

(i)   Highlight cells C2 to C13; to do this, click with the left-hand mouse button on cell C2, hold down that button and drag the mouse down to cell C13, then release the button.

(ii) Click on the Excel's Home tab,   Home   , then click ⬇ Fill ▾ , then ⬇ Down .

---

[6] To make a cell 'active', click on the cell using a mouse or other pointing device.

[7] When you type **=ln(** into cell C2, Excel anticipates the function you want to use and promptly shows LN(number). Excel has many built-in functions, and such anticipation can save you the effort of fully typing a function.

[8] The number that appears in cell C2 is given to 7 figures. In fact, Excel holds numbers internally to 15 figures. These extra figures are useful for limiting the effect of rounding errors when calculations are performed. However, we must remember to give final answers to the appropriate number of significant figures, as discussed in Section 2.5.

Sheet 8.4 Calculation of the natural logarithm of the number of counts in cell C2

	A	B	C
1	Thickness (mm)	Counts	ln(Counts)
2	1	1213	7.100852
3	2	981	

Sheet 8.5 Use of FILL DOWN command

	A	B	C
1	Thickness (mm)	Counts	ln(Counts)
2	1	1 213	7.100852
3	2	981	6.888572
4	3	752	6.622736
5	4	631	6.447306
6	5	423	6.047372
7	6	491	6.196444
8	7	329	5.796058
9	8	259	5.556828
10	9	215	5.370638
11	10	158	5.062595
12	11	141	4.94876
13	12	118	4.770685

Within a fraction of a second the logarithms are calculated and appear in the C column, as shown in Sheet 8.5.

Suppose that, having completed the calculations, we find that something has been overlooked: before taking the natural logarithm of the counts we should have subtracted the contribution to the counts due to the background gamma radiation. Making the necessary corrections would require a fair amount of extra work if we were to do this by hand; however, the steps required are easily performed using a spreadsheet. We begin by typing **ln(Counts-background)** into cell D1, as shown in Sheet 8.6. The background was 42 counts and we type **=LN(B2-42)** into cell D2.

As before, when the Enter key is pressed the calculation is carried out. The result is shown in Sheet 8.7.

Sheet 8.6 Column allowing background subtraction before the taking of logs

	A	B	C	D
1	Thickness (mm)	Counts	ln(Counts)	ln(Counts-background)
2	1	1 213	7.100852	= LN(B2-42)
3	2	981	6.888572	

Sheet 8.7 Setting up cell D2 to calculate ln(Counts-background)

	A	B	C	D
1	Thickness (mm)	Counts	ln(Counts)	ln(Counts-background)
2	1	1 213	7.100852	7.065613
3	2	981	6.888572	

Sheet 8.8 FILL DOWN command used in the D column

	A	B	C	D
1	Thickness (mm)	Counts	ln(Counts)	ln(Counts-background)
2	1	1 213	7.100852	7.065613
3	2	981	6.888572	6.844815
4	3	752	6.622736	6.565265
5	4	631	6.447306	6.378426
6	5	423	6.047372	5.942799
7	6	491	6.196444	6.107023
8	7	329	5.796058	5.659482
9	8	259	5.556828	5.379897
10	9	215	5.370638	5.153292
11	10	158	5.062595	4.75359
12	11	141	4.94876	4.59512
13	12	118	4.770685	4.330733

Using the FILL DOWN command, the function is copied from cell D2 into all the cells down to cell D13, as shown in Sheet 8.8.

Though a spreadsheet does little that cannot be done by hand, laborious and repetitive calculations, which are prone to mistakes, are avoided.

8.3   Built-In Statistical Functions

As well as the natural logarithm function used in Section 8.2.1, spreadsheets have built into them the mathematical functions that you would normally find on a scientific calculator, such as SIN, COS, TAN and EXP. Statistical functions are available that calculate, for example, the mean and standard deviation of values. To use these functions, you need to indicate the cells that contain the values to which the function is to be applied and specify the cell where the outcome of the calculation should appear.

8.3.1   Worked Example Using Built-In Statistical Functions

Sheet 8.9 shows an example using spreadsheet functions. A total of 77 values representing the breaking strength (in newtons) of a batch of carbon fibres are shown. Beneath the columns of numbers are the values of the maximum and minimum breaking strengths and the mean and standard deviation[9] of the values contained in the cells A2 through G12. Each of these quantities is calculated by specifying the range of cells containing the values to be included in the calculation (rows 2 through 12 and columns A through G) and the statistical function to be applied. The specific functions employed in Sheet 8.9 are shown in Sheet 8.10.[10]

  Although a spreadsheet is normally arranged as rows and columns of equally sized cells, the layout of the spreadsheet can be modified to make it easier to read. In Sheet 8.9 a large heading has been added to improve the clarity of the spreadsheet. It makes sense to take advantage of these presentation options, as a sheet of cells filled only with numbers, formulae and functions can be difficult to comprehend later.

Exercise A

The CORREL function in Excel calculates the linear correlation coefficient, r, for x–y data.

  Enter the data in Table 8.1 into Excel. Use the CORREL function to calculate the linear correlation coefficient of the data.

Table 8.1 x–y data for Exercise A

x	42	78	94	102	150	178	202	230	255
y	423	398	380	258	224	198	167	140	122

---

[9] The standard deviation calculated here is the estimate of the population standard deviation, s, given by equation 5.7.

[10] The mean and standard deviation in cells E14 and E15 have been rounded to the nearest whole number.

Sheet 8.9 Example of use of some of Excel's built-in statistical functions

	A	B	C	D	E	F	G
1				Strength of carbon fibres (N)			
2	107	87	153	117	104	158	100
3	81	126	99	112	125	110	114
4	189	166	113	185	142	108	125
5	145	117	168	137	97	120	139
6	139	133	151	134	110	121	89
7	110	119	108	117	125	130	126
8	189	89	125	154	102	107	101
9	135	155	178	115	145	158	126
10	109	128	77	112	129	111	142
11	126	104	122	126	81	116	110
12	140	180	107	98	131	147	136
13							
14	max =	189		mean =	126		
15	min =	77		std. dev =	25		

Sheet 8.10 Built-in functions used in Sheet 8.9

	A	B	C	D	E
14	max =	=MAX(A2:G12)		mean =	=AVERAGE(A2:G12)
15	min =	=MIN(A2:G12)		std. dev =	=STDEV.S(A2:G12)

8.3.2 The LINEST Function

LINEST is a versatile function that can determine, for example, the slope and intercept of the best line fitted to x–y data using linear least squares.[11] Owing to the widespread use of least squares in analysing experimental data, we will consider this function in some detail.[12]

LINEST differs from other Excel functions, such as AVERAGE or CORREL, in that it is an *array* function. Array functions carry out a series of calculations on the

---

[11] Note that the LINEST function cannot perform weighted least squares.

[12] Another way to fit a line to x–y data using Excel is to first plot the data and then use Add Trendline. For details, see https://www.excel-easy.com/examples/trendline.html, or Kirkup (2012), chapter 6.

Table 8.2 Voltage and current values for a semiconductor device									
Voltage, V (mV)	5	10	15	20	25	30	35	40	45	50
Current, I (mA)	12	15	13	11	8	4	4	2	1	2

contents of an array of cells and return the results of those calculations in another array of cells. For example, LINEST can calculate the slope, intercept and the standard errors in slope and intercept, and return these in an array of cells.[13]

Table 8.2 shows values that have been gathered in an experiment to investigate the electrical characteristics of a semiconductor device.[14] As the voltage applied to the device increases, so the current changes.

The relationship between the current, I, and the voltage, V, for the device can be written

$$I = DV \exp\left(-\frac{V}{B}\right), \tag{8.1}$$

where D and B are constants to be estimated using the method of least squares.

We first linearise equation 8.1 using the methods discussed in Section 3.3.5. To linearise equation 8.1, divide both sides by V, then take natural logarithms to give

$$\ln\left(\frac{I}{V}\right) = -\frac{V}{B} + \ln(D). \tag{8.2}$$

This equation is of the form $y = mx + c$, where $y = \ln(I/V)$ and $x = V$. The slope, m, and intercept, c, are given by

$$m = -\frac{1}{B} \tag{8.3}$$

$$c = \ln(D). \tag{8.4}$$

It follows that

$$B = -\frac{1}{m} \tag{8.5}$$

$$D = \exp(c). \tag{8.6}$$

We now apply the LINEST function to the data in Table 8.2.

Sheet 8.11 shows a spreadsheet incorporating the LINEST function. Equation 8.2 indicates that $\ln(I/V)$ is required. Values of $\ln(I/V)$ are calculated in the C column.

---

[13] LINEST can calculate other quantities, but we will not consider those here. See Kirkup (2012), chapter 7, for more information on the LINEST function.

[14] The device in the experiment is a tunnel diode.

Sheet 8.11 Use of the LINEST function

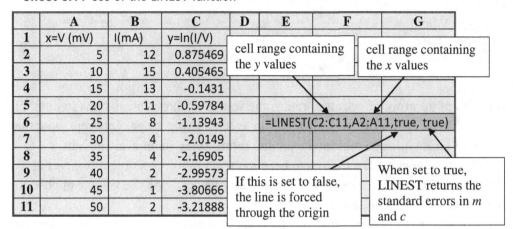

	A	B	C	D	E	F	G
1	x=V (mV)	I(mA)	y=ln(I/V)				
2	5	12	0.875469				
3	10	15	0.405465				
4	15	13	-0.1431				
5	20	11	-0.59784				
6	25	8	-1.13943		=LINEST(C2:C11,A2:A11,true, true)		
7	30	4	-2.0149				
8	35	4	-2.16905				
9	40	2	-2.99573				
10	45	1	-3.80666				
11	50	2	-3.21888				

cell range containing the y values

cell range containing the x values

If this is set to false, the line is forced through the origin

When set to true, LINEST returns the standard errors in m and c

Sheet 8.12 Values returned after pressing Ctrl-Shift-Enter

	A	B	C	D	E	F	G
1	x=V (mV)	I(mA)	y=ln(I/V)				
2	5	12	0.875469				
3	10	15	0.405465				
4	15	13	-0.1431				
5	20	11	-0.59784				
6	25	8	-1.13943		-0.10447	1.392409	
7	30	4	-2.0149		0.006835	0.212046	
8	35	4	-2.16905				
9	40	2	-2.99573				
10	45	1	-3.80666				
11	50	2	-3.21888				

m

c

s_m

s_c

A pointing device, such as a mouse, is used to highlight the four cells E6 to F7. It is into these cells that Excel returns the results of its calculations.

To have Excel return the results of its calculations into cells E6 to F7, we hold the Ctrl and Shift keys down together, then press the Enter key. Once this is done, the spreadsheet looks like Sheet 8.12.[15]

---

[15] We have added annotation to Sheet 8.12 to indicate which cells contain the slope, intercept and their respective standard errors.

Taking s_m and s_c to be the uncertainties in m and c, respectively, we write

$$m = (-0.104 \pm 0.007)\ (\text{mV})^{-1}$$

$$c = (1.4 \pm 0.2).$$

Using equations 8.5 and 8.6, we can determine B and D:

$$B = -\frac{1}{m} = -\frac{1}{-0.104\ (\text{mV})^{-1}} = 9.57\ \text{mV}$$

$$D = \exp(c)\ = \exp(1.4) = 4.02.$$

Exercise B

(1) Construct the spreadsheet shown in Sheet 8.11.
(2) Use the standard errors in m and c as shown in Sheet 8.12 to determine the uncertainties in B and D.
(3) Change the value in cell B2 from 12 to 10. What are the new values for m and c and the standard errors in these values?

8.4 Visualising Data Using a Spreadsheet

Despite the importance of data entered into a spreadsheet, their presentation as rows and columns of numbers is visually dull, such that interesting features such as outliers or gaps in the data may go unrecognised. It is better to have a pictorial representation of data. Spreadsheets offer a range of options for presenting data in pictorial form.

8.4.1 The Histogram

A popular way to display data such as that appearing in Sheet 8.9 is to create a histogram, allowing you to examine the distribution of data – for example, are the values spread evenly or do some values occurs more frequently than others? One approach to creating a histogram using Excel is to:

(i) place the values to be displayed as a histogram into a single column of numbers in Excel
(ii) highlight the values to be displayed
(iii) insert a histogram by going to the top of the Excel screen and clicking on Insert , then ▮▮▮ ▾, then ▮▮▮ .

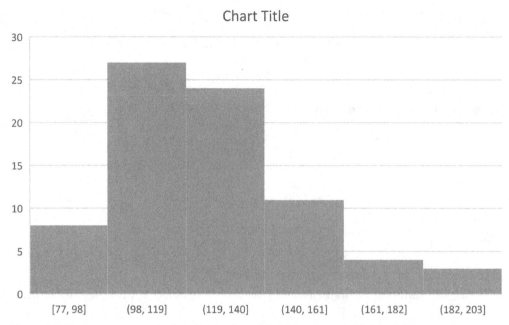

Figure 8.2 Histogram, prepared using data in Sheet 8.9, of the strength of carbon fibres.

Figure 8.2 shows the values in Sheet 8.9 presented in the form of a histogram created using Excel. The spreadsheet has ordered the values and counted the number occurring in each interval (the interval is often referred to as a 'bin') appearing along the x axis. As created, the histogram is unsatisfactory, as it lacks a title, labelled axes or an indication of units. By clicking on the chart in Excel, these important features can be added to the histogram.[16]

8.4.2   The x–y Graph

If in your experiment you have gathered, say, 50 pairs of x–y values, the prospect of plotting them by hand on graph paper can be discouraging. A spreadsheet can handle thousands of data points and plot them in under a second. With a click of a button, the data and graph can be transferred to a printer to generate a paper copy of the work. Alternatively, the graph of the data can be 'cut and pasted' into a report of the experiment.

As an example of an Excel graph containing many data, Figure 8.3 shows data acquired during an experiment which studied the output voltage of a thermoelectric generator[17] as a function of time. The data, consisting of 600 pairs of voltage and time values, were transferred from the recording instrument to an Excel spreadsheet. This allowed the data to be plotted, examined and stored in an Excel file.

---

[16] See https://www.excel-easy.com/data-analysis/charts.html for more information about adding essential details to Excel charts.

[17] A thermoelectric generator is a device that converts heat into electrical energy.

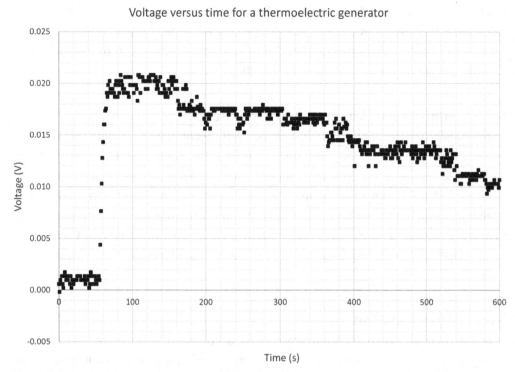

Figure 8.3 Output voltage versus time graph for a thermoelectric generator exposed to a heat source at time = 60 s.

8.5 Other Features Offered by Excel

Although this chapter has introduced spreadsheets mainly with regard to their use in analysing and presenting data from experiments, the scope of their features extends well beyond this. Options offered by sophisticated packages such as Excel include the manipulation of complex numbers, data smoothing and Fourier analysis. Here we will describe some other features common to spreadsheets and useful for the analysis of experimental data.

8.5.1 'What If' Calculations

A 'what if' calculation is no more than an application of the techniques we have already discussed for the modification of data in spreadsheets. Suppose we require the answers to the following questions:

In a linear least-squares problem, how are the slope, m, and intercept, c, affected (and the uncertainties in these quantities) if the uncertainties in the individual y data values

   (i)   increase?
   (ii) decrease?
   (iii) are proportional to the magnitude of the y quantity?

Additionally, what happens to m and c if an outlier is added or removed?

These questions could take a considerable amount of work to answer using a hand calculator. However, once a spreadsheet is created to perform a weighted least-squares fit to data, values can be changed or removed or the uncertainties modified, and the consequences examined immediately.

8.5.2   Transferring Data into a Spreadsheet

Spreadsheets can accept data that have been recorded by a data acquisition system. Some experiments generate so much data, such as the data shown in Figure 8.3, that it is unrealistic to record them by hand. In such circumstances, data are logged by an instrument. The data generated are stored in a file, then transferred to a spreadsheet for analysis and presentation. A common way of storing data in a form readable by programs such as Excel is as a CSV file.[18] The process of transferring data into a program is usually referred to as 'importing' a file.

8.5.3   Excel's Analysis ToolPak

It is possible to assemble a spreadsheet to accomplish a range of calculation-intensive analyses. Recognising that some types of analysis are common, Excel has an add-in program, called the Analysis ToolPak, to help with these.[19] The ToolPak comprises tools that allow for the numerical analysis of data, and in some cases offers graph-plotting options. A limitation of most tools in the ToolPak is that when data are changed, calculations using those data are *not* automatically updated, requiring the tool to be re-run. Nevertheless, for some applications, the ToolPak is very convenient. The ToolPak can be found at the top of the Excel screen under the Data tab, Data . By clicking on Data Analysis , a dialog box appears which reveals the tools that are available.

Table 8.3 summarises some of the analysis tools available in Analysis ToolPak.

8.5.4   Non-Linear Least Squares

A powerful facility that spreadsheets offer is the ability to solve problems in situations where an analytical solution is not possible. When an equation is not linear in the unknown parameters, for example

$$y = a \, \exp\left(bx\right) + c, \tag{8.7}$$

linearisation is not possible and so linear least squares cannot be used to estimate the parameters a, b and c. However, using a spreadsheet, a, b and c can be estimated

---

[18] CSV stands for comma separated values.

[19] A detailed introduction to the Analysis ToolPak, including how to install it, can be found at https://www.excel-easy.com/data-analysis/analysis-toolpak.html

Table 8.3 Summary of some tools in Excel's Analysis ToolPak

Tool	What it does
Descriptive Statistics	Calculates several statistics including the mean, standard deviation and range of the data.
Histogram	Calculates the number of values within each bin of the histogram and offers the option to plot the histogram. The bin size can be calculated by Excel or chosen by the user. The latter is generally preferable, as Excel's own choice generally results in peculiar bin intervals.
t-Test Two Sample Assuming Equal Variances	There are situations where the means of two sets of data need to be compared. This might occur, for example, where the mercury contamination at two locations in an estuary is of interest. A hypothesis might be proposed that the means of data gathered at each location are the same, and a statistical test of this hypothesis is required.[a] Such a comparison needs to take into account the spread of data obtained at each location. The t-Test tool can make this comparison so long as the standard deviations of the data from all locations are approximately the same.
Random Number Generation	Generates random numbers based on several underlying probability distributions, such as the normal or Poisson distribution. The tool is useful, for example, when studying the impact of different amounts of variability in x–y data on the parameter estimates emerging from fitting by least squares, as well as on the size of the standard errors in these estimates. More generally, this tool is useful in many computer simulations.
Regression[b]	Fits equations to data using least squares, where the equation is linear in the parameters to be estimated. Examples of equations that can fitted are $y = mx + c$, $y = ax^2 + bx + c$, and $y = ax + b/x + c$. The standard error in each parameter estimate is also calculated, as well as other useful statistics.[c] There are options for graphing the data.

[a] See Kirkup (2012), chapter 9, for an introduction to hypothesis testing using Excel.
[b] 'Regression' is a term commonly used to refer to fitting an equation to data by linear least squares.
[c] See Liengme (2015), chapter 8.

iteratively. The technique involves setting up the sum of squares, $\sum (y_i - \hat{y}_i)^2$, in a cell in the spreadsheet and then instructing the spreadsheet to minimise that sum by varying the adjustable parameters (which for equation 8.7 would be the parameters a, b and c).

Solver is an add-in in Excel capable of finding the best estimates of parameters using least squares.[20] Solver can be configured so that it can vary the parameters until

---

[20] See Kirkup (2012), chapter 8, for examples of the use of Solver.

the sum of squares is minimised and then returns their final values. In principle this facility allows us to extend least-squares fitting to take account of any function.[21]

8.6   Alternatives to Spreadsheets

Spreadsheets are attractive because of their ease of use, extensive on-line help and availability. As they have been designed to accommodate a wide range of users such as those from business and commerce, not all the features that scientists and engineers need are to be found in a spreadsheet. There are packages designed specifically for the analysis and presentation of scientific and technical data which offer extra options to those found on spreadsheets, such as numerical differentiation and integration of plotted data, factor analysis, cross-tabulation tables and non-parametric tests. Where extra data analysis features are required, there are many powerful packages on the market designed for scientists and engineers, such as Origin.[22] Nevertheless, for many scientists and engineers, a spreadsheet is the first option they turn to when carrying out a preliminary analysis of data.

8.7   Comment

Pencil, paper and a scientific calculator are often adequate when analysing small amounts of data. However, spreadsheets are favoured, especially when large amounts of data are involved.

Some care is required when using a spreadsheet. To use a spreadsheet most effectively requires a degree of organisation. Spreadsheet files can easily be mislaid if they are not carefully catalogued. Stored data can become corrupted (or even subject to computer viruses), so care must be taken to ensure that files are backed up.

Despite these minor drawbacks, the power, flexibility and ease of use of a spreadsheet makes it an excellent tool for the analysis and presentation of data in science and engineering.

---

[21] This type of fitting is referred to as non-linear regression. See Bates and Watts (2007) for more information.

[22] For details of Origin, go to https://www.originlab.com/Origin.

9 | Computer-Aided Data Capture

9.1 Why Use a Computer to Assist in Data Capture?

As well as being powerful tools for data analysis, computers routinely play a major role in capturing data generated during experiments. The benefits of using a computer to capture data include the ability to:

(i) record changes in quantities spanning short to long time intervals (say, between microseconds and months)
(ii) collect data simultaneously from several sources
(iii) display data in tabular or graphical form in real time
(iv) store large amounts of data in a form readable by a data analysis program
(v) gather data without requiring the continuous attention of the experimenter
(vi) control aspects of an experiment, such as adjusting the output voltage from a power supply or starting and stopping a motor.

In this chapter we consider interfacing and data gathering using a computer.

Interfacing is a general term used to describe the process of connecting external devices to a computer so that data from the devices can be captured, transferred, stored and analysed.

9.1.1 Computers Handle Numbers

Quantities measured during an experiment must be represented within a computer as numbers. In an experiment we might want to measure quantities such as temperature, pressure, pH or humidity. How can such quantities be transformed into numbers?

There are basically three stages. The first stage is to use a transducer to produce a voltage (or in some cases a current) proportional to the quantity being measured.[1] The second stage is to modify the voltage, for example if it is too large or too small, so that it is suitable for delivery to the third stage. Voltages produced by transducers

---

[1] The term 'sensor' is sometimes used instead of 'transducer'. As both terms are in common usage we will use both in this chapter.

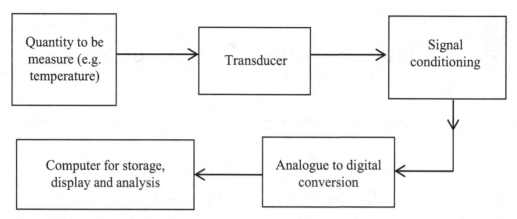

Figure 9.1 Block diagram of a typical computer-based data gathering system.

are often termed *signals* and the modification process is called *signal conditioning*. The third stage is to use a device called an *analogue to digital converter*,[2] or ADC for short, to convert a voltage to a number. After a number has been generated, it is transferred to the memory of a computer ready for storage, display and analysis.

A typical system for gathering data is shown in the form of a block diagram in Figure 9.1.

The transducer, signal conditioner and ADC may be quite separate components in a data gathering system. However, a device may comprise these plus other elements such as a microcontroller.[3] Such a device is sometimes referred to as a 'smart sensor'.[4] The microcontroller within the smart sensor supports communication with other devices such as a computer.

9.2   Transducers

To monitor a physical quantity with a computer, we need a means of generating a voltage (or current) proportional to the quantity, such as pressure or light intensity, being studied. This process is often referred to as 'converting' the quantity to a voltage.

Table 9.1 contains examples of transducers along with the typical change in the signal voltage or current you can expect for a given change in the physical quantity.

Examining the last column in Table 9.1 we see that some transducers generate voltages in the millivolt range and others in the volt range. Similarly, some

---

[2] The term 'analogue' voltage refers to the fact that, within certain limits, say 0 to 5 V, a voltage can take on any value.

[3] A microcontroller is basically a computer on a single chip.

[4] For an introduction to smart sensors, see Meijer et al. (2014).

Physical quantity	Transducer	Typical measurement range	Typical change in signal voltage or current
temperature	thermocouple	0 to 100 °C	0 to 5 mV
pressure	piezoelectric material	0 to 3×10^5 Pa	0 to 250 mV
relative humidity	capacitor	5 to 100%	1 to 4 V
light intensity	photodiode	10 to 10 000 Lx	0.1 μA to 1 mA
magnetic field	Hall effect probe	0 to 40 mT	2 to 6 V
position or displacement	precision potentiometer	0 to 200 mm	0 to 10 mV

Table 9.1 Commonly used transducers

transducers generate currents spanning microamps to milliamps. It is vital that a measurement system cope with such large ranges of voltage and current. In an experiment in which a multimeter is used to measure voltages and currents, the problem is easily dealt with: by simply turning a dial or pressing a button on the multimeter, we can select a range which allows us to measure the voltage or current with the desired resolution. When using an ADC, we need to find a way to match the output voltage or current from the transducer to the input of the ADC.

9.3 Signal Conditioning: A Little Electronics Goes a Long Way

ADCs permit the best resolution of a quantity being measured when the output range of the signal-conditioning circuitry of the data gathering system is matched to the input range of the ADC. A typical input range for an ADC is 0 to 5 V. Suppose that, during an experiment, a transducer generates a voltage that varies in the range 0 to 70 mV. How can we convert 0 to 70 mV to 0 to 5 V?

The answer is that we need an amplifier which has an 'amplifying factor', more usually termed *voltage gain*, of

$$\frac{5 \text{ V}}{70 \text{ mV}} = 71 \quad \text{(to two significant figures)}.$$

A cost-effective way of providing the necessary voltage gain is to use a device called an *operational amplifier*.

9.3.1 The Operational Amplifier

An operational amplifier, or 'op-amp' for short, is a multipurpose electronic device adaptable to many situations requiring signal conditioning. Connections to an

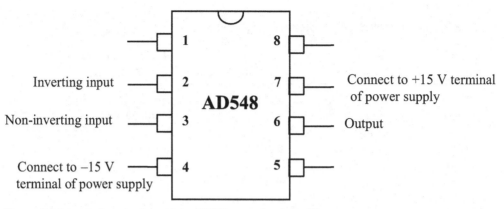

Figure 9.2 Layout of connections of the AD548 operational amplifier.

op-amp are made through terminals often referred to as 'pins'. Figure 9.2 shows the layout of the pins of the AD548 op-amp.[5] Pins 1 and 5 are used to provide small adjustments to the output of the op-amp,[6] and we will not deal with their use here.

Op-amps have two inputs and one output. Many operate from a power supply which provides voltages that are positive and negative with respect to its zero volts terminal.

It is important that the transducer output should not be affected by being attached to the amplifier. For example, if we attach an amplifier to a transducer and this causes the voltage from the transducer to decrease from 70 mV to 60 mV, then the amplifier is having an adverse effect on the output of the transducer and the circuit requires modification.

For the output voltage of a transducer to remain unaffected when it is connected to signal-conditioning circuitry, the input resistance of the circuitry must be several orders of magnitude larger than the output resistance of the transducer. The AD548 op-amp shown in Figure 9.2 has an input resistance of about 10^{13} Ω, which is large enough for all but the most demanding of applications.

9.3.2 Example of Use of an Op-amp

We describe how an op-amp such as the AD548 can be used to provide a gain of about 70 to a signal from a transducer.

The AD548, in common with many other op-amps, is often operated using a power supply capable of providing $+15$ V and -15 V (dc) with respect to the zero volts terminal of the power supply.[7]

Pins 1, 5 and 8 do not play a part in the circuit in Figure 9.3 and are left without any connection. V_{in} is the voltage from the transducer and V_{out} is the voltage after

---

[5] The AD548 is manufactured by Analog Devices, Norwood, MA.

[6] Pin 8 has no electrical function.

[7] The current required to operate an op-amp is typically of the order of tens of milliamps. Most power supplies are capable of providing this.

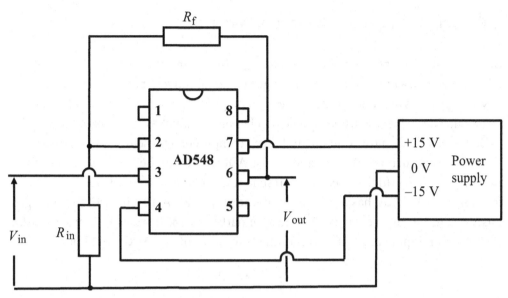

Figure 9.3 Op-amp connected as an amplifier.

amplification has occurred. Two resistors, R_f and R_{in}, set the voltage gain of the amplifier. The relationship between V_{in} and V_{out} is[8]

$$V_{out} = \left(1 + \frac{R_f}{R_{in}}\right) V_{in}. \tag{9.1}$$

To complete the design, if we choose R_f to be 83 kΩ and R_{in} to be 1.2 kΩ, then the voltage gain, which is defined as

$$\text{voltage gain} = \frac{V_{out}}{V_{in}},$$

would be equal to (by rearranging equation 9.1)

$$\text{voltage gain} = \left(1 + \frac{R_f}{R_{in}}\right) = 1 + \frac{83 \text{ kΩ}}{1.2 \text{ kΩ}} \approx 70$$

(gain has no units). The gain of about 70 means that a voltage from a transducer of, say, 30 mV, will be increased by the amplifier to about 2.1 V.

It is worth noting that op-amps are versatile devices capable of more than amplifying a voltage. For example, with the addition of capacitors, an op-amp can act as a filter which reduces the influence of unwanted signals that interfere with the output of the transducer.

---

[8] For a detailed discussion of op-amps and their applications, see Carter and Mancini (2017).

9.4 The Analogue to Digital Converter

Before a computer can store or analyse voltages from a transducer, the voltage must be transformed into a number using an analogue to digital converter (ADC). An ADC accepts any voltage on its input side within a specified range (such as 0 to 5 V or 0 to 10 V) and on its output side delivers a positive whole number proportional to the input voltage.

Because, at their most fundamental level, computers represent numbers in binary, we hear terms such as 10-, 12- and 16-bit ADCs. The number of 'bits' refers to the number of binary digits used to represent the voltage converted by the ADC. In general, an n-bit ADC has a resolution of one part in 2^n. A 10-bit ADC, for example, has a resolution of one part in 2^{10}, i.e. one part in 1024. If we were to use a 10-bit ADC with an input range of 0 to 5 V, the resolution of the ADC would be

$$\frac{5 \text{ V}}{1024} = 4.88 \text{ mV}.$$

Similarly, a 12-bit ADC has a resolution of one part in 2^{12}, i.e. one part in 4096. If this ADC has an input range of 0 to 10 V, its resolution is

$$\frac{10 \text{ V}}{4096} = 2.44 \text{ mV}.$$

The greater the number of bits of the ADC, the higher is its resolution. Common types of ADC available have 12-bit resolution and can typically complete a conversion in under 50 μs, allowing 20 000 conversions or more to be made each second. At a cost, there are ADCs with more bits and shorter conversion times.

ADCs are generally available as part of a data acquisition system. Such a system normally comprises more than ADCs, and includes, for example, circuitry for timing and counting events.

Exercise A

A particular 16-bit ADC can convert voltages in the range 0 to 1.25 V. What is the resolution of this ADC?

9.5 Data Acquisition, Processing and Analysis Options

Computer-based data capture and analysis systems generally comprise a data acquisition system (DAQ) supported by software that can carry out a multitude of functions. Such software supports the creation of files containing experimental data readable by software packages like Excel, and the display of the data, and offers statistical analysis options. Systems with the capacity to acquire and analyse data vary in ease of use, versatility and sophistication.

Some systems are 'plug and play', such that they automatically recognise the type of transducer connected to a computer. Once the transducer has been connected, the computer is ready to collect and display data from the transducer. Other data gathering systems offer the potential for a high degree of customisation necessary for the control of complex experiments, and can gather data from many transducers simultaneously.

9.5.1 Plug and Play Data Acquisition and Analysis System

Plug and play systems allow you to use a desktop or laptop computer for data acquisition and analysis, and do not require detailed knowledge of sensors, signal conditioning or interfacing.

As an example, Logger Lite[9] allows for plug and play data acquisition using software and sensors made by Vernier.[10] Once the program is downloaded, installed and running, the software automatically recognises the type of sensor. A sensor is connected, via an interface made by Vernier, to a USB port on a computer. As soon as the connections are made, an empty table and a graph are displayed, ready to be populated with data. Data acquisition commences once the user clicks the 'Collect' ▶ button.

Figure 9.4 shows a screen from an experiment using Vernier hardware and software in which the voltages from two thermoelectric generators were measured using voltage sensors.[11] The toolbar at the top of the screen gives access to a range of useful commands. Those include options to:

- change the rate at which data are gathered
- choose the total duration over which data are gathered
- auto-scale the axes as data collection proceeds
- fit a straight line to data using least squares
- export data in a form readable by a spreadsheet or statistical analysis program.

The use of plug and play sensors supported by user-friendly software and hardware is a quick and convenient way to make measurements. However, this approach limits you to the sensors designed for use with the plug and play system. An experiment may require the use of sensors not supported by the software, or

---

[9] Logger Lite is a free program for PCs and Mac computers downloadable from https://www.vernier.com/products/software/logger-lite/

[10] Vernier Software & Technology of Beaverton, OR, has developed hardware and software suitable for use in colleges and universities. PASCO of Roseville, CA, is another company supplying software and plug and play sensors suitable for data capture and analysis.

[11] Other plug and play sensors that can be used with Logger Lite include thermometers, accelerometers and pressure sensors.

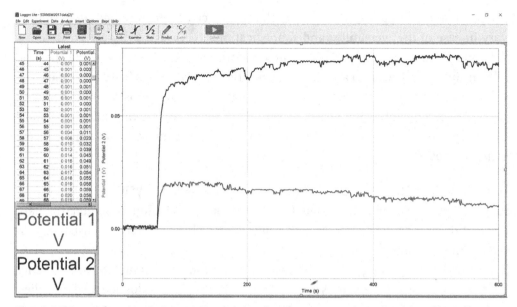

Figure 9.4 Screen from Vernier Logger Lite showing data from an experiment involving thermoelectric generators.

high-resolution data acquisition beyond the capability of the available devices. In such situations it is possible to adopt a more versatile approach to data capture which employs a microcontroller such as the Arduino.[12] We consider the Arduino next.

9.5.2 Low-Cost, Versatile Data Acquisition and Analysis System: The Arduino

The Arduino Uno (AU)[13] is a versatile, low-cost device suited to assisting with data capture and the control of experiments. The AU has been designed to be used by experimenters with widely varying backgrounds, from those with little or no background in electronics and computer programming to those with an extensive knowledge of both.

The AU consists of a small circuit board, as shown in Figure 9.5, that can be connected to a computer through one of the computer's USB ports. This allows the AU to be programmed for data gathering tasks.[14] The AU can output digital signals[15] to external devices, allowing it, for example, to switch motors on and

---

[12] The Arduino (see https://www.arduino.cc/) is a popular microcontroller. Many companies, such as Vernier, who supply data capture and control hardware and software have chosen to incorporate this microcontroller as one of their data gathering options.

[13] The Arduino is available in many versions, including the Arduino Uno, Arduino Leonardo and Arduino Mega.

[14] The program can be stored in the AU, allowing the program to run without being connected to the computer.

[15] A digital signal takes on one of two values, typically 0 V or 5 V.

Figure 9.5 Arduino circuit board.

off or open and close valves. The AU is also able to take a digital signal from an external device, enabling it to establish the state of the device. As examples, it can verify whether a switch is open or closed or an LED is off or on. The AU also includes an ADC comprising six analogue input channels which can accept voltages in the range 0–5V. Each channel provides 10-bit resolution and is capable of making up to 10 000 measurements per second.

The software and hardware for the AU is open-source, meaning that:

- anyone can design and make Arduino-compatible hardware
- AU computer code is available publicly for free to modify or adapt. This also allows programmers to collaborate on enhancements and additions to the code.

Programs written for the AU vary from the simple to the complex. As examples, a program may simply switch an LED on and off. A more advanced application would see an Arduino used to control a drone in flight. A typical application of the Arduino for data capture would be to make measurements of the voltage output of a transducer such as a thermocouple.[16] Figure 9.6 shows a program for the

---

[16] Note that the voltage from a thermocouple would be too small to measure directly with an AU, such that some signal conditioning would be required.

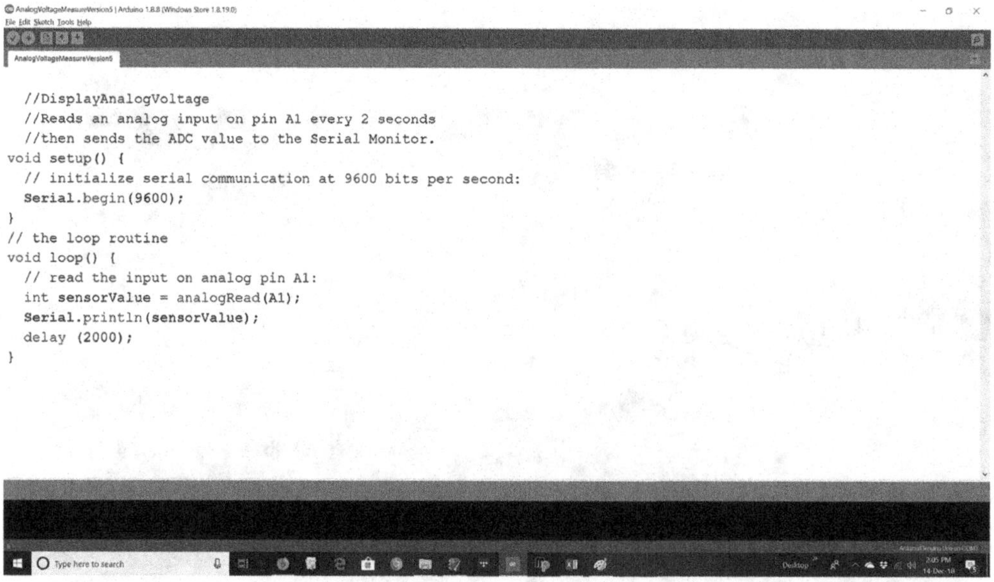

Figure 9.6 Example of a program for the AU.

Arduino which takes a voltage (which could be from a transducer) every 2 s and converts it to a number.

The language used to program the AU is based on the C programming language. A common approach to developing a program for making measurements using the AU is to find an existing working program (often from the Internet) which does much of what you require, then to adapt it to your specific needs. This requires that an amount of experience be gained in writing programs for the AU. Introduction to programming the AU can be found in texts and on the Internet.[17]

The versatility and cost-effectiveness of the AU make it a popular choice for control and data gathering from experiments using a computer. Another option is to employ ready-made robust hardware and software supplied by companies such as LabJack[18] which specialise in computer-based data capture and analysis systems.

9.5.3 Data Gathering with Smartphones

The convergence of computer, communication, display and sensing technologies has led to the development of smartphones capable of measuring physical quantities. As examples, some smartphones have built-in magnetic field, humidity and pressure sensors. The portability of the sensors allows for experiments to be carried out far from a laboratory.

---

[17] For a good book which introduces programming, see Monk (2016). Another source of useful information on programming can be found at http://www.me.umn.edu/courses/me2011/arduino/arduinoGuide.pdf

[18] LabJack Corporation, CO, https://labjack.com/

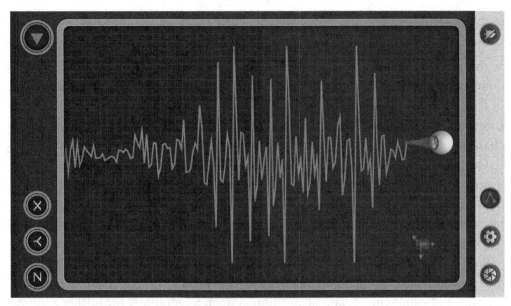

Figure 9.7 Screenshot of the seismometer tool in Skypaw's Toolbox.

Applications (or 'apps') are available for smartphones, usually at small cost, which exploit the on-board sensors and allow data from the sensors to be stored and transferred, for example by a file attached to an email, so that the data can be analysed by a statistics package.

Figure 9.7 shows an example of an app that uses a smartphone's built-in sensors. The screenshot in Figure 9.7 is from the seismometer tool contained in Skypaw's Toolbox, available for iPhones.[19] Similar apps are available for Android smartphones.

An attraction of smartphones for data gathering, in addition to their portability, is their capacity to communicate wirelessly over short ranges via Bluetooth.[20] Many companies manufacture compact sensors for data gathering using Bluetooth.[21]

A challenge, as with all measurement systems, is to ensure that values obtained from the sensors are not misleading. For example, though a smartphone sensor might show the size of a magnetic field, without checking the calibration of the sensor, the values obtained cannot be assumed to be accurate.

---

[19] http://skypaw.com/toolbox.html

[20] Bluetooth is a communication technology, first developed in the mid 1990s, which allows for two-way communication between devices such as smartphones and transducers. In addition to smartphones, the technology is routinely found on PCs, Macs, tablets and laptops.

[21] An example of such a company is BlueMaestro, https://www.bluemaestro.com/our-difference-bluetooth/

9.6   Comment

In this chapter we have considered the application of computers to data gathering. Computers play a key role in data gathering as well as the analysis and presentation of data derived from experiments. An appreciation of the power and limitations of computers used in this context requires some familiarity with the performance and characteristics of transducers, signal-conditioning circuits and DAQ software.

The convenience of computer-based DAQ systems can sometimes lead an experimenter to take the data gathered at face value. All sensors/transducers have their limitations of range, resolution, speed of response, as well as accuracy. Just as measurements made without the aid of a computer need to be scrutinised for sources of random and systematic error, so must measurements made with a computer.

Books that consider data gathering using computers, as well as instruction manuals provided by manufacturers of DAQ systems, are of great help when creating and employing a data acquisition system. Suggestions for further reading at the end of this book may be of assistance if you wish to devise your own computer-based data gathering system.

APPENDIX 1
Degrees of Freedom and the *t* Distribution

Degrees of Freedom

The estimate of the population standard deviation, s, is given by

$$s = \left(\frac{\sum (x_i - \bar{x})^2}{n - 1} \right)^{\frac{1}{2}}. \tag{5.7}$$

$n - 1$ in the denominator of the equation is equal to the number of *degrees of freedom* in the calculation (the symbol v – Greek nu – is usually used to represent this quantity). The reason why v is 1 less than the number of data is that we have placed a small restriction on the values that x_i can take. This arises from the fact that, in order to calculate the population standard deviation, we first need to know the population mean, and the best we can do is to use the mean of the sample of data, $\bar{x}$, as an estimate of the population mean. So, for example, if the mean of five numbers is 7.3 and four of the numbers are 8.7, 5.3, 6.2 and 7.9, the last number *must* be 8.4. In this situation, one degree of freedom has been 'lost' because, although (in this example) four of the five numbers can have any value, the fifth cannot.

A similar situation arises when estimating the population standard deviation, s, of x–y data that lie about a straight line drawn through the data. In Chapter 6 the equation for s was given by

$$s = \left[\frac{1}{n - 2} \sum (y_i - m x_i - c)^2 \right]^{\frac{1}{2}}. \tag{6.14}$$

Here the number of degrees of freedom is $n - 2$, as *two* restrictions are placed on the possible values of y because two quantities are calculated using the x–y data, namely the slope, m, and the intercept, c.

In general, whenever a population parameter, such as the population standard deviation, is estimated using sample data drawn from that population, one degree of freedom is lost for every quantity appearing in the equation which is calculated using the sample data.

The *t* Distribution

When we calculate the confidence interval for the true value of a quantity based on a sample drawn from a population (as we do when we make repeat measurements of a quantity), the characteristic width of the normal distribution, given by the standard deviation, is an underestimation of the spread in the data values. In fact, the smaller the sample (i.e. the smaller the number of repeat measurements), the larger is the spread.

The statistical distribution we use to describe the spread of small samples of data is called the *t* distribution (or, sometimes, Student's *t* distribution). It is very similar to the normal distribution in that it is a symmetrical distribution which 'tails away' as the deviation of a value from the mean increases. Figure A1.1 shows the normal distribution and the *t* distribution.

The *t* curve is flatter than the normal curve and there is more area in the tails beneath the curve. For both curves the area beneath the curve between any two values of x (e.g. x_1 and x_2 in Figure A1.1) is proportional to the probability that a value of x will be observed within that interval. We can see that the area between x_1 and x_2 under the curve is greater for the *t* curve than for the normal curve. This means that, where the *t* distribution is applicable, there is a greater probability that a value x will be observed in this interval (which is far from the mean) in comparison to the normal distribution.

Whereas the shape of the normal curve is independent of the number of data points, n, the *t* curve becomes flatter as the number of points decreases. For large values of n the *t* curve and the normal curve are indistinguishable.

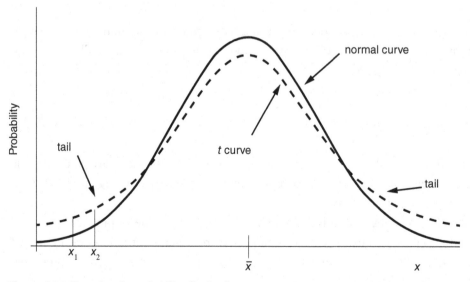

Figure A1.1 Normal and *t* probability distributions.

Table A1.1 Values of t_c at the 95% confidence level for between 4 and 30 measurements

Number of data points, n	Number of degrees of freedom, v	Values of t_c for the 95% confidence level
4	3	3.18
8	7	2.36
12	11	2.20
20	19	2.09
30	29	2.05

In situations where the number of repeat measurements is small, the confidence interval for the true value, μ, of a quantity can be written

$$\bar{x} - t_c s_{\bar{x}} < \mu < \bar{x} + t_c s_{\bar{x}}, \tag{A1.1}$$

This can be written more succinctly as $\mu = \bar{x} \pm t_c s_{\bar{x}}$. where $s_{\bar{x}} = s/\sqrt{n}$. s is given by equation 5.7. t_c, referred to as the critical value of the t distribution, depends on the number of data points and the level of confidence required.

Textbooks concentrating on the statistics of data, such as that by Meyer (1975), contain tables of t_c for a wide range of confidence levels and number of data points. Table A1.1 shows a small table for the 95% confidence level, which is probably the most widely used level.

As an example, suppose we want the 95% confidence interval for the true value of the following repeat measurements of the mass of a body (units kg): 6.5, 8.0, 7.4, 7.2. Using these data, we have

$\bar{x} = 7.275\,\text{kg}$
$s = 0.6185\,\text{kg}$
$s_{\bar{x}} = 0.3093\,\text{kg}$

Using equation A1.1 and referring to Table A1.1, we see that for $n = 4$ the 95% confidence level value of t_c is 3.18, so that

$$\text{true value} = (7.275 \pm 3.18 \times 0.3093)\,\text{kg}$$
$$= (7 \pm 1)\,\text{kg}.$$

If we had used the normal distribution, we would have found the 95% confidence interval to be $(7.3 \pm 0.6)\,\text{kg}$. When quoting confidence intervals to one significant figure the difference between using the normal and t distribution is usually negligible when n exceeds 10.

APPENDIX 2

Propagation of Uncertainties Where Errors Are Uncorrelated

Suppose a quantity V depends on a and b so that we can write $V = V(a,b)$. If the error in the value of a does not correlate with the error in the value of b, we can determine the uncertainty in V using the ideas about quantifying the variability in data discussed in Chapter 5.

We begin by writing the variance of V as (see equation 5.1)

$$\sigma_V^2 = \frac{\sum(V_i - \bar{V})^2}{n} = \frac{(\Delta V_i)^2}{n}. \qquad (A2.1)$$

As V depends upon a and b, we can write

$$\Delta V_i = \frac{\partial V}{\partial a}\Delta a_i + \frac{\partial V}{\partial b}\Delta b_i.$$

Substituting this into equation A2.1 gives

$$\sigma_V^2 = \frac{\sum\left(\frac{\partial V}{\partial a}\Delta a_i + \frac{\partial V}{\partial b}\Delta b_i\right)^2}{n}. \qquad (A2.2)$$

The right-hand side of equation A2.2 can be expanded to give

$$\sigma_V^2 = \sum\frac{\left(\frac{\partial V}{\partial a}\Delta a_i\right)^2}{n} + \sum\frac{\left(\frac{\partial V}{\partial b}\Delta b_i\right)^2}{n} + \sum\frac{2\frac{\partial V}{\partial a}\frac{\partial V}{\partial b}\Delta a_i\Delta b_i}{n}. \qquad (A2.3)$$

The first two terms on the right-hand side of equation A2.3 will always be greater than zero and so cannot be neglected. By contrast, the product $\Delta a_i\Delta b_i$ in the third term will sometimes be positive and at other times negative, depending on the signs of Δa_i and Δb_i. To a good approximation, the last term in equation A2.3 will be zero, with positive terms in the summation cancelling out the negative terms. We are then left with

$$\sigma_V^2 = \left(\frac{\partial V}{\partial a}\right)^2\sum\frac{(\Delta a_i)^2}{n} + \left(\frac{\partial V}{\partial b}\right)^2\sum\frac{(\Delta b_i)^2}{n}.$$

We can write

$$\sigma_a^2 = \sum\frac{(\Delta a_i)^2}{n} \quad \text{and} \quad \sigma_b^2 = \sum\frac{(\Delta b_i)^2}{n},$$

so that

$$\sigma_V^2 = \left(\frac{\partial V}{\partial a}\right)^2 \sigma_a^2 + \left(\frac{\partial V}{\partial b}\right)^2 \sigma_b^2. \tag{A2.4}$$

We have shown that, for measurements in which the errors in values obtained by measurement are uncorrelated, the variance in the calculated value is equal to the sum of the variances in the measured quantities.

When random errors are dominant in our measurements, it is the standard error of the mean of a set of repeated measurements that we take to be the uncertainty in a value obtained through measurement. In order to calculate the standard error in $\bar{V}$, $s_{\bar{V}}$, when the standard errors in $\bar{a}$ and $\bar{b}$ are given by $s_{\bar{a}}$ and $s_{\bar{b}}$, respectively, σ_V in equation A2.4 is replaced by $s_{\bar{V}}$, σ_a by $s_{\bar{a}}$ and so on. Equation A2.4 now becomes

$$s_{\bar{V}}^2 = \left(\frac{\partial V}{\partial a}\right)^2 s_{\bar{a}}^2 + \left(\frac{\partial V}{\partial b}\right)^2 s_{\bar{b}}^2. \tag{A2.5}$$

Therefore, the standard error in $\bar{V}$, $s_{\bar{V}}$, can be written

$$s_{\bar{V}} = \sqrt{\left(\frac{\partial V}{\partial a}\right)^2 s_{\bar{a}}^2 + \left(\frac{\partial V}{\partial b}\right)^2 s_{\bar{b}}^2}. \tag{A2.6}$$

Solutions to Exercises and Problems

Chapter 2

Exercises

A

(i) 2.2 μV, (ii) 62 mm, (iii) 65.2 kJ, (iv) 0.18 MW, (v) 67 pF, (vi) 82 nPa, (vii) 0.144 mH, (viii) 167 kg

B

Pressure ($\times 10^5$ Pa)	1.03	1.01	1.01	0.99	1.05	1.08

C

(i) three, (ii) two, (iii) three, (iv) four, (v) three, (vi) five, (vii) five, (viii) three, (ix) three

D

(i) 18.9, (ii) 0.108, (iii) 725, (iv) 1.76, (v) 62 700, (vi) 0.00725, (vii) 42 000, (viii) 1060

E

(i) 10, (ii) 562, (iii) 0.00024, (iv) 469, (v) 0.185, (vi) 71.4, (vii) 6.46, (viii) 101, (ix) −171

F

(1) (i) 5.654×10^{-3}, (ii) 1.250×10^2, (iii) 9.384×10^7, (iv) 3.400×10^6, (v) 1.001×10^{-7}, (vi) 5.724×10^{-1}, (vii) 6.377×10^3, (viii) 9.544×10^{-3}

(2) (i) 5.7×10^{-3}, (ii) 1.3×10^2, (iii) 9.4×10^7, (iv) 3.4×10^6, (v) 1.0×10^{-7}, (vi) 5.7×10^{-1}, (vii) 6.4×10^3, (viii) 9.5×10^{-3}

G

From the example, the heat required to raise the temperature by 5 °C is $Q \approx 6 \times 10^{10}$ J. $Pt = Q$, where P is the power supplied for a time t. Rearranging gives

$$t = \frac{Q}{P}.$$

Assumption: a domestic kettle supplies power, $P = 2\,\mathrm{kW}$ $(= 2\,\mathrm{kJ/s})$. It follows that

$$t \approx \frac{6 \times 10^{10}\,\mathrm{J}}{2 \times 10^{3}\,\mathrm{J/s}} = 3 \times 10^{7}\,\mathrm{s}$$

$$t\,(\text{days}) \approx \frac{3 \times 10^{7}}{24 \times 60 \times 60} \approx 350 \text{ days},$$

i.e. close to one year!

Assumption: all the heat goes to raising the temperature of the water (i.e. there are no heat losses).

Problems

2.1. 46.9 g

2.2. $5.2 \times 10^{-2}\,\mathrm{kg}$

2.3. The student has (a) inserted the diameter of the sphere into formula for the volume instead of the radius, (b) omitted the units for the density and (c) given the volume to too many significant figures and used incorrect units for the volume. Inserting the radius gives the volume (to three significant figures) as 56.5 mm³. The value for the density of the sphere must be changed, given to the correct number of significant figures (two, if we adopt the rule given in Section 2.5.2) and the units must be inserted. The density of the sphere turns out to be $7.8 \times 10^{-3}\,\mathrm{g/mm^3}$.

2.4. To estimate the volume of my body, I imagined it to be rectangular prism, as shown in Figure S.1. This is a very crude approximation, but probably sufficient to estimate the volume to within a factor of 2 of the actual volume.

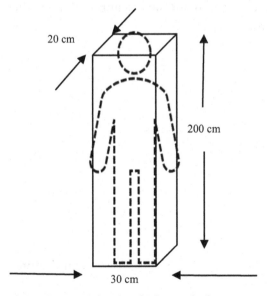

Figure S.1 Approximating the human body as a rectangular prism.

Working to one significant figure, I took the height of the prism, h, to be 200 cm, its breadth, b, to be 30 cm and its width, w, to be 20 cm. I found it easier to estimate the dimensions of the prism in centimetres, rather than in metres or millimetres.

The volume, V, of the prism is $V = hbw = 200 \times 30 \times 20 = 1 \times 10^5 \, \text{cm}^3$ (to one significant figure). So, the answers to parts (i) to (iii) of the question are:

(i) $1 \times 10^8 \, \text{mm}^3$
(ii) $1 \times 10^5 \, \text{cm}^3$
(iii) $1 \times 10^{-1} \, \text{m}^3$

As an alternative approach, I estimated the volume of my body using the relationship $\rho = m/V$, where ρ is the average density of my body, m is my mass and V is my volume. As a human consists mostly of water, it appears reasonable to take the density of water ($1 \, \text{g/cm}^3$) to be my average body density. My mass is about 80 kg, which means

$$V = \frac{m}{\rho} = \frac{80 \times 10^3 \, \text{g}}{1 \, \text{g/cm}^3} = 8 \times 10^4 \, \text{cm}^3.$$

This answer is about 20% less than that obtained using the rectangular prism approximation. When solving a Fermi problem by two methods, a difference of 20% between estimates is regarded as insignificant.

2.5. To find the surface area of my body, I used the rectangular prism approximation (and the dimensions) described in the answer to question 2.4. The prism has six faces. The area of each face is the product of the height of the face multiplied by its width. Add together the areas of all six faces to get the total area. I made the total area to be $2 \times 10^4 \, \text{cm}^2$ (to one significant figure). So, the answers to parts (i) to (iii) of the question are:

(i) $2 \times 10^6 \, \text{mm}^2$
(ii) $2 \times 10^4 \, \text{cm}^2$
(iii) $2 \, \text{m}^2$

Chapter 3

Exercises
A

Faults:

(i)   The title of the graph is incorrect: the graph shows time versus temperature, not temperature versus time.
(ii) The units have been omitted from the x axis.
(iii) The data point gathered at temperature 42 °C has not been plotted.
(iv) The last data point on the graph has been incorrectly plotted.

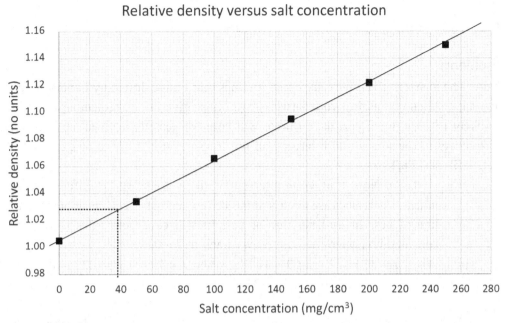

Figure S.2 Relative density versus salt concentration.

B

(1) (i) 0.0045 T, (ii) 0.0090 T

(2) From the line on Figure S.2, when the relative density is 1.030, the corresponding salt concentration is about 38 mg/cm³.

C

My best line through the points gave a slope, m, of 0.026 Ω/°C. I found the coordinates of one point on the line to be $x = 64\,°C$ and $y = 6.85\,Ω$. Using equation 3.6, the intercept (where the temperature is 0 °C) $c = 6.85 − 0.026 \times 64 = 5.2\,Ω$ to two significant figures. The assumption made is that the relation between resistance and temperature remains linear down to 0 °C.

D

Equation number	What to plot	Slope, $m =$	Intercept, $c =$
1	F versus N	μ	0
2	v versus t	a	u
3[a]	R/T versus T	B	A
4	$\ln(I)$ versus x	$-\mu$	$\ln(I_0)$
5[b]	T versus $\sqrt{l}$	$2\pi/\sqrt{g}$	0
6	$1/v$ versus $1/u$	-1	$1/f$
7	H versus T	C	$-CT_0$
8	$\ln(I/V)$ versus V^2	$-B$	$\ln(A)$

[a] An alternative is to plot R/T^2 versus $1/T$. In this case the slope would be A and the intercept would be B.

[b] An alternative is to plot T^2 versus l. In this case the slope would be $4\pi^2/g$. The intercept would still be equal to 0.

E

$\text{Log}_{10}(A) = -10.4$, so $A = 3.9 \times 10^{-11}\,\text{W/K}^4$, $n = 4.0$

Problems

Note: The answers given here are based on lines that I have drawn by hand through x–y data. When comparing your answers with those given here, remember that it is unlikely that the lines you have drawn will coincide exactly with mine, so some differences in the slopes and intercepts of our lines of best fit are to be expected.

3.1. (i) 331 Pa/K

(ii) -182 Pa. Note that your intercept is likely to be quite different from mine. This is because even a small difference between the slope of your line of best fit and my line of best fit will cause our lines to diverge considerably, once they are drawn back to $T = 0\,\text{K}$.

(iii) When $T = 330$ K, $P = 1.09 \times 10^5\,\text{Pa}$; when $T = 400$ K, $P = 1.32 \times 10^5\,\text{Pa}$

(iv) When $P = 1.00 \times 10^5\,\text{Pa}$, $T = 303$ K

3.2.

Equation number	What to plot	Slope, $m =$	Intercept, $c =$
1	T_w versus R^2	$-k$	T_c
2	D versus v^2	$\frac{1}{2}CA\rho$	0
3	$1/f$ versus $1/r$	$n-1$	0
4	E versus f	h	$-\phi$
5	D^2t versus D	AB	A
6	E versus $\cos\theta$	$E_0 v/c$	E_0
7	ΔT versus $\log_{10}A$	k	D

3.3. (i) Take logs of both sides of the equation, so that $\log_{10}T = a\log_{10}d + \log_{10}k$. Plotting $\log_{10}T$ versus $\log_{10}d$ will give a straight line with slope $m = a$ and intercept $c = \log_{10}k$.

(ii) $k = 69$ cm s, $a = -1.00$

(iii) When $d = 45$ cm, $T = 1.53$ s

3.4. (i) $1/v = -1/u + 1/f$

(ii) Plot $1/v$ versus $1/u$; slope $m = -1$, intercept $c = 1/f$

(iii) From my graph, I found the intercept to be $c = 0.054\,\text{cm}^{-1}$, so that $f = 19$ cm, to two significant figures.

3.5. (ii) Slope $m = 6.2$ N/kg, intercept $c = -0.1$ N

(iii) $y = 6.2x - 0.1$

(iv) (a) when mass = 0.70 kg, minimum force = 4.2 N
      (b) when mass = 1.30 kg, minimum force = 8.0 N

(v) Uncertainty in the slope = 0.5 N/kg, uncertainty in the intercept = 0.4 N, so we can write $m = (6.2 \pm 0.5)\,\text{N/kg}$ and $c = (-0.1 \pm 0.4)\,\text{N}$.

3.6. (i) $1/I = R/E + r/E$, so plotting $1/I$ versus R will give a slope $m = 1/E$ and intercept $c = r/E$.

(ii) $E = 1/m$, $r = c/m$

3.7. (iii) Concentration = (1.30 ± 0.15) ppb

3.8. (iii) $m = 1.51$, $c = 1.35$

(iv) $n = 1.51$, $k = 22.4\,\Omega\,A^{-1.51}$

3.9. (i)

Concentration (mol)	Reaction rate (s^{-1})
0.20	0.0400
0.16	0.0333
0.14	0.0313
0.12	0.0250
0.08	0.0179
0.02	0.0086

(iii) Slope of line = $0.174\,\text{mol}^{-1}\,\text{s}^{-1}$

Chapter 4

Exercises

A

(1) I used an electric kettle which shuts off automatically when the water reaches boiling point. I recorded the following times for 600 cm^3 of water to begin to boil:

Time (s)	135	115	119

Before the kettle was switched on the first time, the kettle was at room temperature. After the first boiling the average temperature of the kettle increased and so when the hot water was replaced with water, some heat was transferred from the kettle to the replacement water, even before the kettle was switched on. This reduced the amount of heat required from the element to heat the water to boiling point, thus reducing the time to reach this point.

Other factors that could have affected the heating time are:

(i) variability in the kettle thermostat, i.e. it does not always switch the kettle off at the same temperature every time

(ii) a change in the temperature of the tap water between experiments

(iii) variations in the amount of water added each time.

(2) In an experiment in which a coin was dropped from a distance of 3 m, I recorded the following data:

Time (s)	0.80	0.66	0.80	0.83	0.78	0.82	0.82	0.80	0.84	0.87

The largest value is 0.87 s; the smallest value is 0.66 s.

Synchronising the starting of the watch with the fall of the coin and reacting to the coin hitting the ground are likely to be the most important factors affecting the timing. This could easily introduce an uncertainty of about 0.2 s to any individual measurement.

B
(1) 46 cm³, 3.83×10^{-7} m³/s
(2) 4.3×10^{-3} Pa

C
(1) 0.018 s
(2) Mean = 425 mm, uncertainty = 4 mm

D
(i)   Too many figures in mean *and* in uncertainty: it should be (12.6 ± 0.6) N/m.
(ii)   Too many figures in mean, and it would be better to write this using scientific notation, i.e. $(8.8 \pm 0.5) \times 10^3$ kg/m³.
(iii) No units are given! They should be m/s.
(iv)   The mean should be quoted to the same number of decimal places as the uncertainty, i.e. (1.100 ± 0.001) µF.
(v)   '± 1' does not convey anything meaningful. It *could* mean $\pm 1 \times 10^{-7}$ m, but it is not clear.
(vi)   The power produced is $(9.5 \pm 0.5) \times 10^2$ W.
(vii) The pH is 7.50 ± 0.07.

E
(i)   Fractional uncertainty = 0.1, percentage uncertainty = 10%
(ii)   Fractional uncertainty = 0.04, percentage uncertainty = 4%
(iii) Fractional uncertainty = 0.3, percentage uncertainty = 30%
(iv)   Fractional uncertainty = 0.11, percentage uncertainty = 11%
(v)   Fractional uncertainty = 0.013, percentage uncertainty = 1.3%
(vi)   Fractional uncertainty = 0.05, percentage uncertainty = 5%

F
(1) (i) $\rho = 4.96 \times 10^2$ kg/m³
     (ii) $\rho_{max} = 5.06 \times 10^2$ kg/m³, $\rho_{min} = 4.87 \times 10^2$ kg/m³, so $u_\rho = 9$ kg/m³
(2) $a = 6.0$ m/s², $a_{max} = 6.6$ m/s², $a_{min} = 5.4$ m/s², so $u_a = 0.6$ m/s²
(3) $v = 46$ m/s, $v_{max} = 50$ m/s, $v_{min} = 42$ m/s, so $u_v = 4$ m/s

G

(1) $\partial s/\partial a = \frac{1}{2}t^2$, $\partial s/\partial t = at$

(2) $\partial P/\partial I = 2IR$, $\partial P/\partial R = I^2$

(3) $\partial n/\partial i = \cos i/\sin r$, $\partial n/\partial r = -\sin i \cos r/\sin^2 r$

(4) $\partial v/\partial T = \frac{1}{2}(1/\mu T)^{\frac{1}{2}}$, $\partial v/\partial \mu = -\frac{1}{2}(T/\mu^3)^{\frac{1}{2}}$

(5) $\partial f/\partial u = v^2/(u+v)^2$, $\partial f/\partial v = u^2/(u+v)^2$

(6) $\partial R/\partial \rho = l/A$, $\partial R/\partial l = \rho/A$, $\partial R/\partial A = -\rho l/A^2$

H

(1) $B = (1.61 \pm 0.12) \times 10^{-3}$ T

(2) $j = (1.04 \pm 0.12) \times 10^6$ A/m^2

(3) $f = (12.0 \pm 0.2)$ cm

(4) $N = (7.7 \pm 1.3) \times 10^3$

Problems

4.1.   (i)   Range $= 1 \times 10^{-2}$ Ω

(ii)   Best estimate is the mean $= 9.44 \times 10^{-2}$ Ω, uncertainty in resistance $= 1.3 \times 10^{-3}$ Ω

(iii)   Resistance of mercury sample $= (9.44 \pm 0.13) \times 10^{-2}$ Ω

4.2. $n = (0.49 \pm 0.13)$ mol

4.3.   (i)   Flow rate $= (0.437 \pm 0.013)$ m/s

pH $= 7.04 \pm 0.07$

Temperature $= (10.5 \pm 0.3)\,^\circ$C

Electrical conductivity $= (9.4 \pm 0.5) \times 10^2$ μS/cm

Lead content $= (52 \pm 5)$ ppb

(ii)   The largest fractional uncertainty is in the lead content (fractional uncertainty $= 0.1$).

4.4.   (i) $k = 6.9$ kg/s^2 (note that the spring constant is normally expressed in units of N/m, which is equivalent to kg/s^2)

(ii) $k = (6.9 \pm 1.4)$ kg/s^2

(iii) $k = (6.9 \pm 1.4)$ kg/s^2

4.5. Using the partial differentiation method to determine the uncertainty, $G = (11.4 \pm 0.7)$ dB.

4.6. $\partial z/\partial a = e^b$, $\partial z/\partial b = ae^b$, $z = (1.24 \pm 0.18) \times 10^5$

4.7.

Uncertainty in $\frac{1}{\text{reaction time}}$ (s^{-1})
0.003
0.002
0.002
0.0019
0.001
0.0002

4.8. $\mu_s = (0.97 \pm 0.03)$

4.9. $p = (3.6 \pm 0.3) \times 10^2\,\text{Pa}$

Chapter 5

Exercises

A

(1) $\bar{x} = 24.81$, $\sigma = 1.70$, $\sigma_{\bar{x}} = 0.601$

(2) $\bar{x} = 175.9\,\text{mm}$, variance $= 5.010\,\text{mm}^2$, $\sigma = 2.238\,\text{mm}$, $\sigma_{\bar{x}} = 0.2889\,\text{mm}$. We can write the rebound height as $(175.9 \pm 0.3)\,\text{mm}$.

B

(i)

Interval (mm)	Frequency
$145 \leq x < 150$	2
$150 \leq x < 155$	2
$155 \leq x < 160$	11
$160 \leq x < 165$	13
$165 \leq x < 170$	16
$170 \leq x < 175$	9
$175 \leq x < 180$	6
$180 \leq x < 185$	1

(ii) The data look to be approximately normally distributed.

(iii) There are 38 values within $\pm\sigma$ of the mean; 57 values lie within $\pm 2\sigma$ of the mean. The normal distribution predicts about 70% of values within $\pm\sigma$ of the mean and 95% within $\pm 2\sigma$ of the mean; 70% of 60 is 42 and 95% of 60 is 57.

C

(1) $\bar{x} = 75.67$, $\sigma = 8.219$, $s = 8.585$

(2) (i) $\sigma_{\bar{x}} = 2.37$, $s_{\bar{x}} = 2.48$, (ii) $\sigma_{\bar{x}} = 2$, $s_{\bar{x}} = 2$

D

(1) $5 \times 10^{-4}\,\text{g/mm}^3$

(2) (i) $n = 1.421$

(ii) $\partial n/\partial i = \cos i / \sin r = 0.7875$, $\partial n/\partial t = -\sin i \cos r / \sin^2 r = -1.818$

(iii) $s_{\bar{n}} = \sqrt{(\partial n/\partial i)^2 s_i^2 + (\partial n/\partial r)^2 s_r^2}$

(iv) Note that $1° = 1.745 \times 10^{-2}\,\text{rad}$; $s_{\bar{n}} = 0.0649$

(v) $n = 1.42 \pm 0.06$ (refractive index has no units)

Problems

5.1.   (i) $\sigma = 0.26$, $s = 0.37$

(ii)   Using the first three numbers, $\sigma = 0.22$, $s = 0.26$; using the first four numbers, $\sigma = 0.19$, $s = 0.22$ (same value to one significant figure); using the first five numbers, $\sigma = 0.17$ and $s = 0.20$.

5.2.   (i) $\bar{x} = 9.11$ mg, $s = 0.504$ mg

(ii) $s_{\bar{x}} = 0.145$ mg

(iii) 70% confidence interval is approximately[1] 8.96 mg to 9.25 mg

5.3.   (i) $\bar{x} = 4.99$ mmol/L, $s = 0.203$ mmol/L

(ii) $s_{\bar{x}} = 0.068$ mmol/L

(iii) 95% confidence interval is approximately[2] $\bar{x} - 2s_{\bar{x}}$ to $\bar{x} + 2s_{\bar{x}}$, i.e. 4.85 mmol/L to 5.12 mmol/L

5.4.   (ii)   Difficult to tell – more data required

(iii) $\bar{x} = 1.102$ s, $s = 0.214$ s, $s_{\bar{x}} = 0.0303$ s

(iv)   velocity $= (320 \pm 10)$ m/s

5.5.   95% confidence interval for the true number of particles is 145 to 157

5.6.   Using method for combining uncertainties described in Section 5.5:
$o.d. = (0.8 \pm 0.2)$

5.7.   Using method for combining uncertainties described in Section 5.5:
$V_{in} = (6.3 \pm 0.3)$ mV

5.8. $\eta_f = (1.6 \pm 0.3)$ kg/(m s)

5.9.   (i) $E = 16\pi^2 ML^3 / (T^2 bd^3)$

(ii) $E = 1.09 \times 10^{10}$ N/m^2

(iii) $u_E = \sqrt{(\partial E/\partial L)^2 u_L^2 + (\partial E/\partial T)^2 u_T^2 + (\partial E/\partial b)^2 u_b^2 + (\partial E/\partial d)^2 u_d^2}$
$= 3.15 \times 10^9$ N/m^2

So we can write $E = (1.1 \pm 0.3) \times 10^{10}$ N/m^2.

Note that there is a bit of work to determine the partial derivatives. If you can access a package that will calculate the derivatives for you (such as that offered free by Wolfram at http://www.wolframalpha.com/calculators/derivative-calculator/) this will save you some effort.

5.10. Note that as the constant, k, is given in base units, all the quantities in the calculations need to be expressed in base units (e.g. the height $h = (0.485 \pm 0.001)$ m and the flow rate $Q = (0.38 \pm 0.02) \times 10^{-6}$ m^3/s).

---

[1] Here we have assumed the normal distribution to be valid, but with only 12 values, the t distribution is more appropriate. Using the t distribution, the interval would be 8.96 mg to 9.26 mg. For information on the t distribution, see Appendix 1. Note that the difference is quite small.

[2] Note that with only nine values it would be better to employ the t distribution when calculating the confidence interval. These would turn out to be $\bar{x} - 2.3s_{\bar{x}}$ to $\bar{x} + 2.3s_{\bar{x}}$ (see Appendix 1).

Best estimate of viscosity from the data $= 1.08 \times 10^{-3}$ Pa s
Uncertainty in the best estimate $= 2.3 \times 10^{-4}$ Pa s
We can write $\eta = (1.1 \pm 0.2) \times 10^{-3}$ Pa s.

Chapter 6

Exercises
A
(1) (i) $m = 0.050$
     (ii) $m = 0.023$
(2) $m = 7.315$, $c = 30.70$
(3) (i) $m = -2.873 \times 10^{-6}$ s^{-2}, $c = 9.794$ m/s^2
     (ii) Multiply out the brackets to give $g' = g - 2gd/R_E$. Comparing this with $y = mx + c$, we see that $c = g$ and $m = -2g/R_E$. From this it follows that $g = 9.794$ m/s^2 and $R_E = -2g/m = 6.818 \times 10^6$ m. Incidentally, this compares with the accepted value[3] for the average radius of the Earth of $\sim 6.37 \times 10^6$ m.

B
For question (2) of Exercise A, $s_m = 0.223$, $s_c = 2.33$, so $m = (7.3 \pm 0.2)$ and $c = (31 \pm 2)$. For question (3) of Exercise A, $s_m = 9.2 \times 10^{-8}$ s^{-2}, $s_c = 5.7 \times 10^{-3}$ m/s^2, so $m = (-2.87 \pm 0.09) \times 10^{-6}$ s^{-2} and $c = (9.794 \pm 0.006)$ m/s^2.

D
(1) (iii) $m = -0.10049$ s^{-1}, $c = 2.889$
     (iv) $s_m = 0.0015$ s^{-1}, so $m = (-0.1005 \pm 0.0015)$ s^{-1}
           $s_c = 0.014$, so $c = (2.889 \pm 0.014)$
     (v) $\tau = (9.95 \pm 0.15)$ s, $V_0 = (18.0 \pm 0.3)$ V
(2) (ii) $I_0 = (2.88 \pm 0.09) \times 10^3$ counts/s, $\lambda = (1.00 \pm 0.02) \times 10^{-2}$ s^{-1}
(3) (ii) $A = (0.289 \pm 0.008)$ mm^2, $B = (2.41 \pm 0.11)$ mm^2
     (iii) $A = (0.290 \pm 0.013)$ mm^2, $B = (2.39 \pm 0.18)$ mm^2
     (iv) Percentage difference in $A = -0.4\%$, i.e. unweighted A is *less* than weighted A by 0.4%.
     (v) Percentage difference in $B = 0.7\%$, i.e. unweighted B is *larger* than weighted B by 0.7%.

E
$r = -0.992$

---

[3] Kaye and Laby (1995).

Problems

6.1.　(i)　$m = -1.2201$, $c = 78.408$

　　　(ii)　$s_m = 0.084$, $s_c = 2.4$

　　　(iii)　$r = -0.988$

　　We write $m = -1.22 \pm 0.08$ and $c = 78 \pm 2$.

6.2.　(i)　Plot I versus $\cos(2\theta)$. Slope will be equal to $I_{max} - I_{min}$ and the intercept will be equal to I_{min}.

　　　(ii)　$m = 0.8671$ and $c = 0.9383$

　　　(iii)　$s_m = 0.03$ and $s_c = 0.02$, so $m = (0.87 \pm 0.03)$ and $c = (0.94 \pm 0.02)$

　　　(iv)　$I_{min} = 0.94$ and $I_{max} = 1.81$

6.3.　(i)　Squaring both sides gives $M^2 = kt + D$; therefore plotting M^2 against t should give a straight line of slope k and intercept D.

　　　(ii)　The uncertainty in M^2 is $2M\Delta M$ (see Section 4.5.2), so the uncertainty in M^2 is not constant (the uncertainty increases as M increases) and a weighted fit is required.

　　　(iii)　To three significant figures, $m = 61.4$ (mg)^2/h and $c = 13.9$ (mg)^2.

6.4.　Rearrange the equation to give $1/g = 1/(np) + s/n$. This is of the form $y = mx + c$, where $y = 1/g$, $x = 1/p$, $m = 1/n$ and $c = s/n$.

　　$m = 1387$ N/kg, $c = 2204$ m^2/kg

　　$n = 7.21 \times 10^{-4}$ kg/N, $s = 1.59$ m^2/N

6.5.　(i)　Rearranging the equation gives $(V - 1)/\theta = D\theta + B$. Therefore, plotting $(V - 1)/\theta$ against θ should give a straight line with slope D and intercept B.

　　　(ii)　$m = (-1.9 \pm 0.2) \times 10^{-6}$ cm^3/(°C)^2

　　　　　　$c = (3.219 \pm 0.013) \times 10^{-3}$ cm^3/°C

6.6.　(i)　$I = I_0 \exp\left(^{eV}/_{nkT}\right)$, so taking natural logs of both sides gives $\ln(I) = {}^{eV}/_{nkT} + \ln(I_0)$. Plotting $\ln(I)$ against V should give a straight line of slope $^e/_{nkT}$ and intercept $\ln(I_0)$.

　　　(ii)　If $y = \ln(I)$, then $\Delta y = (\partial y/\partial I)\Delta I = (1/I)\Delta I$. As $\Delta I/I = 0.02$, which is constant, an unweighted fit should be used.

　　　(iii)　$m = 25.417$ V^{-1}, $c = -22.115$, so $I_0 = 2.49 \times 10^{-10}$ A and $n = 1.521$

　　　(iv)　$s_{I_0} = 3 \times 10^{-11}$ A, $s_n = 0.011$

6.7.　(i)　$t^2 = d^2/v^2 + 4h^2/v^2$, so plotting t^2 against d^2 gives a slope of $1/v^2$ and intercept of $4h^2/v^2$.

　　　(ii)　$m = 0.3237 \times 10^{-6}$ s^2/m^2, $c = 1179 \times 10^{-6}$ s^2

　　　　　　$s_m = 0.0019 \times 10^{-6}$ s^2/m^2, $s_c = 2 \times 10^{-6}$ s^2

　　　(iii)　$v = 1758$ m/s, $h = 30.2$ m

6.8.   (ii) $A = 1.247\,\text{cm}^2$, $s_A = 0.0198\,\text{cm}^2$, so $A = (1.25 \pm 0.02)\,\text{cm}^2$
       (iii) $m = 0.08379\,\text{m}\Omega/\text{mm}$, $c = 0.7795\,\text{m}\Omega$, $s_m = 0.0023\,\text{m}\Omega/\text{mm}$, $s_c = 0.33\,\text{m}\Omega$,
             so $m = (0.084 \pm 0.002)\,\text{m}\Omega/\text{mm}$, $c = (0.8 \pm 0.3)\,\text{m}\Omega$
       (iv) $\rho = (10.4 \pm 0.3) \times 10^{-6}\,\Omega\,\text{m}$

In this problem it might seem appropriate to fit $y = mx$ rather than $y = mx + c$ to the data, as the intercept should be zero (i.e. in this case when $l = 0$, R should be equal to zero). When fitting $y = mx$ to data, the equation for the best value for m is

$$m = \frac{\sum x_i y_i}{\sum x_i^2}.$$

The standard error in the slope is

$$s_m = \frac{s}{\left(\sum x_i^2\right)^{\frac{1}{2}}},$$

where s is given by

$$s = \left[\frac{1}{n-1}\sum (y_i - mx_i)^2\right]^{\frac{1}{2}}.$$

Using these equations, we find $m = 0.0888\,\text{m}\Omega/\text{mm}$ and $s_m = 0.0012\,\text{m}\Omega/\text{mm}$, so that $m = (0.0888 \pm 0.0012)\,\text{m}\Omega/\text{mm}$. This leads to $\rho = (11.1 \pm 0.2) \times 10^{-6}\,\Omega\,\text{m}$.

Note: In most cases it is advisable *not* to force the line of best fit through the origin, as random and systematic errors conspire to influence the line so that is does not pass through the origin. This encourages us to think carefully as to the reasons why the line of best fit, given by $y = mx + c$, does not pass through the origin (for example, systematic errors could be influencing the values of y).

6.9.   (i) $m = 3.768\,\text{L/mol}$, $c = 0.00843$
       (ii) $s_m = 0.066\,\text{L/mol}$, $s_c = 0.015$, so $m = (3.77 \pm 0.07)\,\text{L/mol}$ and $c = (0.008 \pm 0.015)$. The equation of line is $y = 3.77x + 0.008$, so $x = (y - 0.008)/3.77$.
       (iii)   If $y = 0.825$, then $x = 0.217\,\text{mol/L}$.

Using least squares to determine the uncertainty in x when there is uncertainty in slope and intercept is beyond the scope of this book. This is discussed in Kirkup (2012), chapter 6.

6.10. Fitting a line to the data gives slope $= 1.995 \times 10^{-3}\,\text{L/IU}$, intercept $= -3.64 \times 10^{-3}$. When absorbance is 0.337, activity is 171 IU/L, which is outside the normal range.

Chapter 7

Exercises

A

Title

Title is clear and informative. The author's affiliation should be included.

Abstract

Abstract is concise and includes a good summary of the quantitative findings. The first two sentences add little and could be removed (or perhaps transferred to the Introduction).

Introduction

Suitable background is included. The references could have been situated in the text closer to the application mentioned. For example, '... including isolating buildings from the effect of earthquakes (Parvin and Ma, 2001), in automobile suspensions to improve the comfort of the ride for passengers (Yamada, 2007)...'.

Materials and Methods

Enough detail is included to allow another person to repeat the experiment. Few details are given about the spring other than it is 20 cm long. If the part number or the manufacturer is known, then this should be stated in this section. There is no mention of the material from which the spring is made. Measurements were made to check repeatability of values obtained using Method B, but not Method A. The reason for not examining repeatability for Method A could be mentioned here, or in the Discussion section.

Results

Data are well presented in both tables and graphs. As the data are adequately presented as graphs, the tables of data could be transferred to an appendix.

Was the equation $y = bx$ or $y = bx + a$ fitted to both sets of data?

An unweighted fit was done when fitting a straight line to data in Table 1 which appears to be reasonable, as the uncertainty in the values plotted on the y axis is constant. However, the contents of Table 2 indicates that the uncertainty in T^2 is not constant, which suggests that a weighted fit is required (note that the Excel function, LINEST, is not suited to weighted fitting).

Line of best fit should be drawn back to the y axis (for both Figures 2 and 3), so that the intercept can be estimated from the graph.

The intercept of the line of best fit in Figure 3 does not pass close to the origin – this deserves an explanation.[4]

Units for g should be m/s^2.

---

[4] A possible explanation is that the mass of the holder has not been accounted for, such that equation 3 should be written $T = 2\pi\sqrt{(m + M)/k}$, where M is the mass of the holder. It follows that $T^2 = 4\pi^2 m/k + 4\pi^2 M/k$. Plotting T^2 versus m should give a straight line with an intercept of $4\pi^2 M/k$.

Discussion

There is a thorough discussion of factors influencing determination of the spring constant using Method B. A similar discussion of factors affecting measurements made using Method A is absent.

A possible cause for the difference between values obtained by each method is suggested in the Discussion section, which could lead to a follow-up experiment examining the cause.

The Discussion section mentions the '95% confidence intervals' but the values of these intervals are not shown, nor are they mentioned elsewhere in the report.

Conclusion

The conclusion is short and reiterates the main findings of the experiment, i.e. the value obtained for the spring constant using each method and which method is preferred.

References

The references conform to the Harvard referencing style. The Walker reference is missing the year of publication.

Acknowledgements

No acknowledgements have been included in the report. If the work was done in collaboration with other people, then they should be mentioned in the Acknowledgements section.

B

Strengths of the poster in Figure 7.5:

The layout of the poster is simple but effective.
Data are presented clearly in tabular and graphical forms.
Each section of the poster is concise and easy to read.

Suggestions for improvement:

The title should be followed by details of the author(s).
The Discussion section remarks that '... g obtained here is consistent with that published elsewhere ...' but doesn't say what that value of g is.
The first two sentences of the Introduction section could be removed as they add little.

Chapter 8

Exercises
A
$r = -0.954$

B

(2) Uncertainty in $B = 0.6\,\text{mV}$; uncertainty in $D = 0.9$

(3) $m = (-0.102 \pm 0.007)\,(\text{mV})^{-1}$; $c = 1.3 \pm 0.2$

Chapter 9

Exercise

A

Resolution $= 19\,\mu\text{V}$

Further Reading

CHAPTER 1: INTRODUCTION TO EXPERIMENTATION

Kanare, M. H. (1985). *Writing the Laboratory Notebook.* Washington, DC: American
    Chemical Society.
This book discusses the principles of scientific record keeping and describes the key role
record keeping plays in science. Advice is given on how to keep a notebook, applicable to
both academic and industrial laboratories. The advice is supported by specific examples.

CHAPTER 2: CHARACTERISTICS OF EXPERIMENTAL DATA

Weinstein, L. and Adam, J. A. (2008). *Guesstimation: Solving the World's Problems on the
    Back of a Cocktail Napkin.* Princeton: Princeton University Press.
This entertaining introduction to estimation contains many examples from science and
engineering.

CHAPTER 3: GRAPHICAL PRESENTATION OF DATA

Cleveland, W. S. (2004). *The Elements of Graphing Data.* Summit, NJ: Hobart Press.
This well-established book considers graphical presentation, with an emphasis on how to
create graphs in science and engineering that are informative and easy to understand. The
principles of graph construction are considered. Many examples are included of good, and
not-so-good, graphs.

CHAPTER 4: DEALING WITH UNCERTAINTIES

Kirkup, L. and Frenkel, R. (2006). *An Introduction to Uncertainty in Measurement.*
    Cambridge: Cambridge University Press.
This book offers an introduction to international guidelines on calculating and expressing
uncertainty which have been adopted by those working in industry as well those in research
and analytical laboratories.

CHAPTER 5: STATISTICAL APPROACH TO VARIABILITY IN MEASUREMENTS

Bevington, P. R. and Robinson, D. K. (2003). *Data Reduction and Error Analysis for the
    Physical Sciences.* New York: McGraw-Hill.

This highly regarded book includes consideration of probability distributions that underpin error analysis in science. The book's treatment of errors and uncertainty is at a greater depth than we covered in Chapter 5.

CHAPTER 6: FITTING A LINE TO *X–Y* DATA USING THE METHOD OF LEAST SQUARES

Davis, M. P. and Dunn, P. F. (2018). *Measurement and Data Analysis for Engineering and Science*. 4th edn. Boca Raton, FL: CRC Press.
This book adopts an advanced approach to fitting using least squares. The book also covers many other topics of value to those carrying out experiments, particularly for students of engineering. These include consideration of measurement systems, Fourier transforms and digital signal analysis.

CHAPTER 7: REPORT WRITING AND PRESENTATIONS

Zeegers, P., Deller-Evans, K., Egege, S. and Klinger, C. (2011). *Essential Skills for Science and Technology*. Melbourne: Oxford University Press.
Covers communication in science and engineering. It offers advice on report writing, presentation skills and the preparation of posters. It also considers the use of references in reports.

CHAPTER 8: USING EXCEL TO PRESENT AND ANALYSE DATA

Walkenbach, J. (2015). *Excel 2016 Bible*. New York: John Wiley & Sons.
As the title implies, this is a comprehensive reference book which offers extensive coverage of Excel and its capabilities. It has been designed with the diverse needs of beginners and advanced users in mind.

CHAPTER 9: COMPUTER-AIDED DATA CAPTURE

Horowitz, P. and Hill, W. (2015). *The Art of Electronics*. 3rd edn. Cambridge: Cambridge University Press.
Considered by many to be *the* book on all aspects of analogue and digital electronics, it is well written and contains many examples of applications of electronics.

Scherz, P. and Monk, S. (2016). *Practical Electronics for Inventors*. 4th edn. New York: McGraw-Hill Education.
This text covers the basic principles of electricity and electronics, progressively guiding the reader to a practical understanding of modern electronics. A chapter focusing on sensors is included, as is a chapter giving practical advice on how to build and test circuits.

References

Bates, D. M. and Watts, D. G. (2007). *Nonlinear Regression Analysis and its Applications*. New York: John Wiley & Sons.

Berendsen, H. J. C. (2011). *A Student's Guide to Data and Error Analysis*. Cambridge: Cambridge University Press.

Bevington, P. R. and Robinson, D. K. (2003). *Data Reduction and Error Analysis for the Physical Sciences*. New York: McGraw-Hill.

Carter, B. and Mancini, R. (2017). *Op Amps for Everyone*. 5th edn. Oxford: Newnes.

Castelvecchi, D. and Witze, A. (2016). 'Einstein's gravitational waves found at last', *Nature*. DOI: 10.1038/ nature.2016.19361.

Kaye, G. W. C. and Laby, T. H. (1995). *Tables of Physical and Chemical Constants*. http://www.kayelaby.npl.co.uk/ general_physics/2_3/2_3_6.html

Kirkup, L. (2012). *Data Analysis for Physical Scientists: Featuring Excel*. 2nd edn. Cambridge: Cambridge University Press.

Liengme, B. V. (2015). *A Guide to Microsoft Excel 2013 for Scientists and Engineers*. London: Academic Press.

Lyon, A. J. (1980). 'Rapid statistical methods: Part 1', *Physics Education*, 15, pp. 78–83.

Macdonald, J. R. and Thompson, W. J. (1992). 'Least-squares fitting when both variables contain errors: Pitfalls and possibilities', *American Journal of Physics*, 60, pp. 66–73.

Meijer, G., Makinwa, K. and Pertijs, M. (2014). *Smart Sensor Systems: Emerging Technologies and Applications*. New York: John Wiley & Sons.

Meyer, S. L. (1975). *Data Analysis for Scientists and Engineers*. New York: John Wiley & Sons.

Monk, S. (2016). *Programming Arduino: Getting Started with Sketches*. 2nd edn. New York: McGraw-Hill.

Pears, R. and Shields, G. (2010). *Cite Them Right: The Essential Referencing Guide*. 10th edn. New York: Red Globe Press.

Taylor, J. R. (1997). *An Introduction to Error Analysis*. 2nd edn. Sausalito, CA: University Science Books.

Walker, J. (2014). *Fundamentals of Physics*. 10th edn. New York: John Wiley & Sons.

Index